高温材料プロセスにおける
物質移動の基礎とケーススタディー

Metallurgical Mass Transfer
from a lecture note of Professors Brimacombe and Samarasekera

竹内栄一 *Eiichi Takeuchi*　田中敏宏 *Toshihiro Tanaka*

大阪大学出版会

はじめに

　資源・環境・エネルギーという社会問題は、今世紀に入って世界的規模で急速な広がりを見せている。今まで生産性や品質など経済面での競争にリソースを重点化していた産業界も、存在意義や活動を持続的なものとするために、環境エミッションやリサイクルといった"静脈的"課題に対しても積極的に取り組み、製品やその製造プロセスのありさまも、大きく変わりつつある。この中で(1)使用する資源の領域をより拡大し、(2)エネルギー使用量を削減すると共に、(3)環境エミッションを抑制するという新たな課題を解決しつつ、(4)社会・マーケットが必要とする先進素材の製造を経済優位性を保ちながら推進していくために、「Material Processing（素材を製造するプロセス）」を取り扱う研究領域に対して大きな期待がかけられつつある。1970～80年代に製造プロセスの最適化を目標に欧米を中心に展開された「Process Metallurgy」という研究手法は、輸送現象論（熱の移動、物質の移動、運動量の移動（粘性に関係））とメタラジー（化学反応、物理化学、凝固学など）とを融合させたものであるが、前述した背景を受けて、グローバルな規模でさらなる進化の重要性が認識されるところである。本書はこの「Process Metallurgy」の一部を成す「物質の移動と反応が絡んだ現象の解析」を「Metallurgical Mass Transfer」と題し、入門にあたっての理論と応用についてわかりやすく紹介するものである。

　輸送現象論の参考書としては"Transport Phenomena Second Edition" Bird, Stewart, Lightfoot[1]、『輸送現象』水科、荻野[2]などが広く知られているが、これらは物理現象に軸を置いたものである。また、材料系学生に向けた同分野の参考書としては"Rate Phenomena in Process Metallurgy" Szekely, Themelis[3]、『材料工学のための移動現象論』谷口、八木[4]などがあり、素材製造分野における輸送現象解析の重要性を紹介しているものの、メタラジーとのかかわり方についてもう一歩踏み込んだ説明や事例の紹介が期待されるところである。

　本書は、材料を製造するプロセスにおける輸送現象を「物質の移動」に絞り、メタラジー（反応平衡や反応速度など）との関連付けを心掛けると共に、材料製造プロセスにおける解析例を併せて紹介しながら、学生に対しては大学で学ぶ基礎理論が材料製造プロセス開発へどのように活用されているのかを示すと共に、企業技術者・研究者に対しては材料製造プロセスの解析や設計を行うための導入の参考となるよう企画したものである。なお、本書は数式を含む英文と和文から構成されているが、英文および演習問題は筆者が以前ブリティッシュコロンビア大学在学中に受講した講義（J. K. Brimacombe 教授および I. V. Samarasekera 教授）のノートの記述に基づいている。ノート使用の許諾に深く感謝する。

2015年7月30日

著者

Contents

はじめに   i

本書の構成   iv

I. Introduction of Mass Transfer in Metallurgical Process ·· 1

   1. Examples of metallurgical process   1

      1.1 Blast furnace   1

      1.2 Basic oxygen furnace   2

      1.3 Copper converter   2

      1.4 Imperial Smelting furnace   3

   2. Basic equations to describe the metallurgical mass transfer   6

      2.1 Mass balance equation   6

      2.2 Molar flux equation   6

      2.3 Chemical reaction related to mass transfer   7

II. Steady State Diffusion (Fick's 1st law) and Relating Mass Transfer ························ 11

   1. Molecular diffusion and Fick's 1st law including the effect of convection   11

   2. Steady state diffusion problems   16

      2.1 Diffusion of component $\langle 1 \rangle$ in the stagnant medium $\langle 2 \rangle$ 16

      2.2 Diffusion accompanied by a fast heterogeneous chemical reaction   19

      2.3 Diffusion accoumpanied by a homogeneous chemical reaction   24

III. Unsteady State Diffusion (Fick's 2nd law) and Relating Mass Transfer ···················· 29

   1. Fick's 2nd law   29

   2. Mass transfer coefficient and diffusion coefficient   33

      2.1 Film theory   34

      2.2 Penetration theory   35

IV. General Formulation of Mass Transfer in Batch Process ·· 41

   1. Basic equation for batch process   41

   2. Linear thermodynamics   44

   3. Nonlinear thermodynamics   48

   4. The chemical reaction which has an infinity large equilibrium constant   51

V. Application of Mass Transfer Analysis for Continuous Processing　　55

　　1. Differences in general formulation for batch and continuous process　55

　　2. Case study (1); Deoxidation of molten blister copper using a submerged CO jet　57

　　3. Case study (2); Kinetics of vacuum dezincing of lead　65

　　　　3.1. Process description　65

　　　　3.2. Modeling of the continous vacuum zinc distillation　65

VI. Problems and Solutions　　73

　　Problem 1　The diffusion coefficient of zinc vapor in nitrogen　74
　　　　　　　—Solution　76

　　Problem 2　The roasting of zinc sulphide with oxygen　79
　　　　　　　—Solution　81

　　Problem 3　Fume formation in basic oxygen furnace　83
　　　　　　　—Solution　85

　　Problem 4　The electrowinning of zinc　91
　　　　　　　—Solution　94

　　Problem 5　The dissolution rate of tungsten pellets falling through the liquid pool of a continuously cast steel billet　98
　　　　　　　—Solution　101

　　Problem 6　The prediction of decarburization rates in the molten steel experiment　105
　　　　　　　—Solution　108

　　Problem 7　Oxygen absorption in an open-pour stream of continuous casting　112
　　　　　　　—Solution　115

　　Problem 8　Manganese loss during steelmaking　118
　　　　　　　—Solution　120

Additional Short Discussions in Metallurgical Mass Transfer　　124

　　A-1. Decarburization in Q-BOP　124

　　A-2. Evolution of CO in the refining of low carbon ferro-chrome alloy　126

あとがき　128

References　129

Index　130

本書の構成

　本書は大きく分けて、第Ⅰ章、第Ⅱ章、第Ⅲ章での「基礎理論と応用手法の紹介」、第Ⅳ章、第Ⅴ章での「ケーススタディーによる現象の単純化とモデリングの解説」、第Ⅵ章での「演習問題と解答」から成る。

　また、このテキストは英語での授業のノート（英文や数式）をベースに和訳や解説を併記して、本分野の英語表現にも親しめるよう構成したものである。

　下記に、各章の概要と要点を示す。

　Ⅰ章では本書に登場する高温プロセスの概要を紹介する。1節は「製鉄プロセスの高炉と転炉」、「製銅プロセスの転炉（精錬炉）」、および「亜鉛製造プロセス（Imperial Smelting炉）」について概説する。続いて、高温プロセスにおける物質移動現象を記述する方法（物質収支の式、物質流束の式、化学量論の式および反応界面での反応平衡式の適用）について説明する。

- ●プロセスの設備、製造工程の概要
- ●化学反応（反応速度、速度定数など）
- ●高温プロセスにおける反応と物質移動の関連

　Ⅱ章では「定常拡散」の基礎を学ぶ。1節では「Fickの第一法則」の導出、さらに拡散などが引き起こす流れが拡散自体に影響を及ぼす場合の定式化を説明する。2節では「滞留した媒体中の元素の拡散」、「不均一反応を伴う等モル相互拡散および非等モル相互拡散」、「均一反応を伴う拡散」についてのモデル式と解析を紹介する。

- ●Fickの第一法則および流れの影響
- ●滞留している媒体中の成分の拡散
- ●不均一反応を伴う拡散
- ●均一反応を伴う拡散

　Ⅲ章では「非定常拡散」の基礎を学ぶ。1節では「Fickの第二法則」の導出と簡略化について説明する。2節では「境膜説（定常状態）」、3節では「浸透説（非定常状態）」について説明し、それぞれの物質移動係数への「Fickの第二法則」のかかわり方を紹介する。

- Fickの第二法則と簡略化
- 拡散係数と物質移動係数（境膜説）
- 拡散係数と物質移動係数（浸透説）

Ⅳ章では回分（バッチ）プロセスにおいて界面反応を伴う物質移動現象の単純化と定式化を学ぶ。1節では解析方法と基礎式の紹介、2節では液体／液体反応、3節では気体／液体反応、4節では界面で非常に大きい平衡値を有する液体／液体反応における解析について説明する。

- 回分プロセスにおける液／液反応の定式化
- 回分プロセスにおける気／液反応の定式化
- 非常に大きな界面反応の平衡定数を持つ液／液反応の定式化

Ⅴ章では、連続プロセスにおける物質移動の解析手法を学ぶ。1節では回分プロセスと連続プロセスの物質収支の基礎式の違いについて、2節ではケーススタディーとして銅精錬（脱酸）におけるガスジェットに沿った物質移動、3節では亜鉛精錬プロセスにおける減圧下での溶融鉛からの脱亜鉛のモデルを紹介する。

- 回分プロセスと連続プロセスの基礎式の違い
- ケーススタディー（1）銅精錬における脱酸素
- ケーススタディー（2）真空処理による溶融鉛からの脱亜鉛

Ⅵ章では、Ⅰ章～Ⅴ章で学んだ基礎知識を応用し、（1）亜鉛の拡散実験による窒素ガス中の亜鉛蒸気の拡散係数の推定、（2）酸素による硫化亜鉛の焙焼における反応律速過程の予測、（3）転炉におけるフューム生成速度の予測とフューム発生の抑制方法の提案、（4）亜鉛の電解析出における最大電流密度の推定と操業改善の提案、（5）連続鋳造プロセスの凝固終了位置を計測するために使用する放射性同位元素を内蔵したタングステンペレットが溶鋼中を落下する際の溶融予測、（6）酸化性スラグによる溶鋼の脱炭速度の推定、（7）連続鋳造におけるオープン注入時の溶鋼の酸素吸収速度、（8）鋼の精錬時のマンガンロス反応　についての演習を行う。

- 【練習問題と解答のための説明】
    亜鉛の拡散実験による窒素ガス中の亜鉛蒸気の拡散係数の推定
- 【練習問題と解答のための説明】
    酸素による硫化亜鉛の焙焼における反応律速過程の予測
- 【練習問題と解答のための説明】
    転炉におけるフューム生成速度予測とフュームの抑制方法の提案
- 【練習問題と解答のための説明】
    亜鉛の電解析出における最大電流密度の推定と操業改善の提案
- 【練習問題と解答のための説明】
    連鋳プロセスの溶鋼プール中タングステンペレットの溶融予測
- 【練習問題と解答のための説明】
    酸化性スラグによる溶鋼の脱炭速度の推定
- 【練習問題と解答のための説明】
    連続鋳造におけるオープン注入時の溶鋼の酸素吸収速度
- 【練習問題と解答のための説明】
    鋼の精錬時のマンガンロス反応

I. Introduction of Mass Transfer in Metallurgical Process

1. Examples of metallurgical process
1. 1 Blast furnace (BF)

一貫製鉄プロセスは高炉から始まる。最新の大型高炉は高さ100m以上、容積は5000m^{3}を超える巨大な設備である。上部から焼結鉱（主としてFe$_2$O$_3$やCaCO$_3$から成る）およびコークス（石炭を乾留し揮発分を除いて燃焼カロリーを上げ、強度を高めたもの）を層状に挿入する。

下部からは、約1200℃に加熱された空気が供給される。この空気は高炉下部でC（コークス）と反応しCOガスとなって、挿入物中を上昇しながら、酸化鉄を還元していく。

酸化鉄はFe$_2$O$_3$→Fe$_3$O$_4$→FeOの順で還元されると共に融着帯以下ではCと反応してC飽和の銑鉄（iron）となって高炉下部に滴下、滞留する。

焼結鉱に含有されるCaCO$_3$は酸化鉄中に含まれるSiO$_2$と共に溶融酸化物スラグ（slag）となって溶融着帯上部から滴下する。この溶融酸化物は鉄鉱石中に含まれるS等の不純物と反応し、溶銑の純度向上に貢献している。

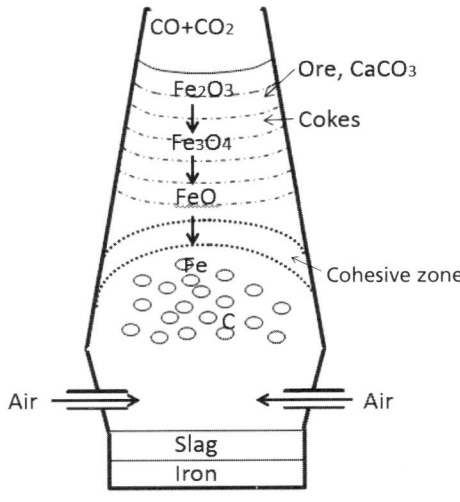

Fig. 1-1　Blast furnace.

$$Fe_2O_3 + CO = Fe_3O_4 + CO_2 \tag{1-1}$$

$$Fe_3O_4 + CO = 3FeO + CO_2 \tag{1-2}$$

$$FeO + CO = Fe + CO_2 \tag{1-3}$$

They are the reactions of heterogeneous and they take place at an interface of between gas/solid. Temperature is at high.

1.2 Basic oxygen furnace (BOF)

高炉で製造された銑鉄は、必要に応じて脱硫（Sを除去する）処理を行った後、転炉に送られる。現在の一般的な転炉では、溶融した銑鉄と共にスクラップ（計約300トン）がCaOなどと共に挿入された後、水冷銅ランスによって純酸素ガスが音速で供給される（たとえば3～4 Nm^3/min・ton-steel）。精錬反応の初期では酸素と反応しやすい溶銑中のSiが除去される（脱珪期）。続いて銑鉄中のCが酸素と反応してCOガスとして除去される（脱炭期）。

COガスはランスから供給される酸素ガスと反応（2次燃焼）しCO$_2$ガスとなる。

一方、CaOは生成したSiO$_2$と共にスラグとなり、溶鋼中のPと反応して溶鋼の純度向上に寄与する。

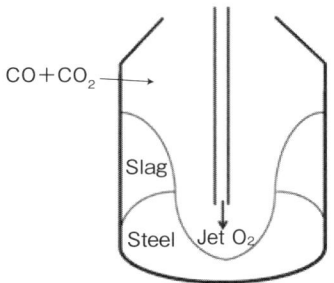

Fig. 1-2　Basic oxygen furnace.

$$\underline{C}+1/2O_2 \rightarrow CO \tag{1-4}$$
$$CO+1/2O_2 \rightarrow CO_2 \tag{1-5}$$

The former reaction is heterogeneous (gas/liq. interface), and the latter is homogeneous (gas/gas). Also they occur at high temperature.

1.3 Copper converter/(Anode furnace)

銅の鉱石は自溶炉によって溶解され、銅を65％程度含有するマットと酸化鉄やSiO$_2$から成るスラグに溶融分離される。Fig.1-3に示すように、転炉ではマットに酸素（純度35～38％）を吹き込み、マットを酸化させてSを取り除くと共に粗銅（Cu99％）を製造する。転炉内で酸素ガスはマット中を上昇する間に銅中のSと反応してSO$_2$ガスとなり、銅の純度を向上させる。転炉は横型回転炉で容量は最大400トンである。その後、粗銅はブタン等のガスによって還元され酸素を除去した後鋳造され、電解精錬を経て純銅が製造される。

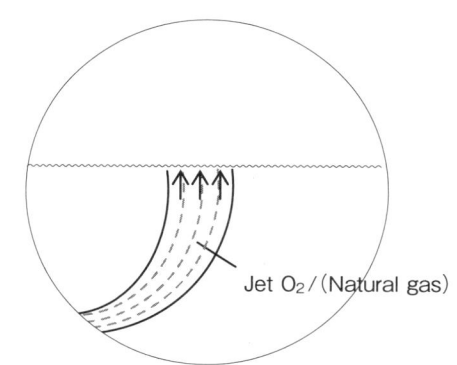

Fig. 1-3　Copper converter/(Anode furnace).

$$Cu_2S+O_2(g) \rightarrow 2Cu+SO_2(g) \tag{1-6}$$

Main reaction occurs at gas/liquid interface at high temperature.

1.4 Imperial Smelting furnace/ (Blast furnace for Zn production)

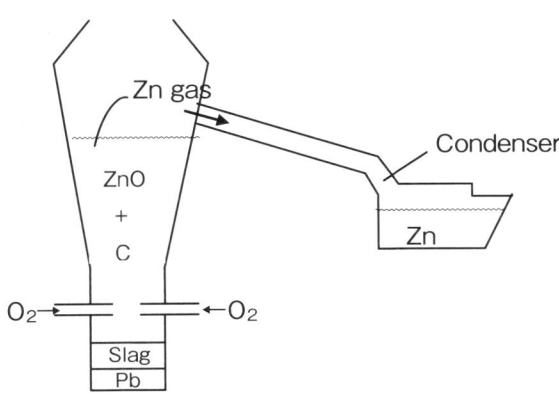

Fig. 1-4 Imperial Smelting furnace.

$$CO + ZnO \rightarrow Zn + CO_2 \qquad (1-7)$$

Main reaction occurs gas / solid interface at high temperature.

Imperial Smelting 社によって開発されたプロセス。焼結した亜鉛と鉛の混合精鉱をコークスと共に炉に挿入する。

炉内に吹き込まれる加熱空気によってCが酸化されCOガスとなって酸化鉛、酸化亜鉛を還元する。

粗鉛は炉底から、亜鉛はガスとして炉頂から取り出され溶融鉛を用いたコンデンサーによって凝集し粗亜鉛となる。

このプロセスにおける反応は主として気体／固体間の不均一反応である。

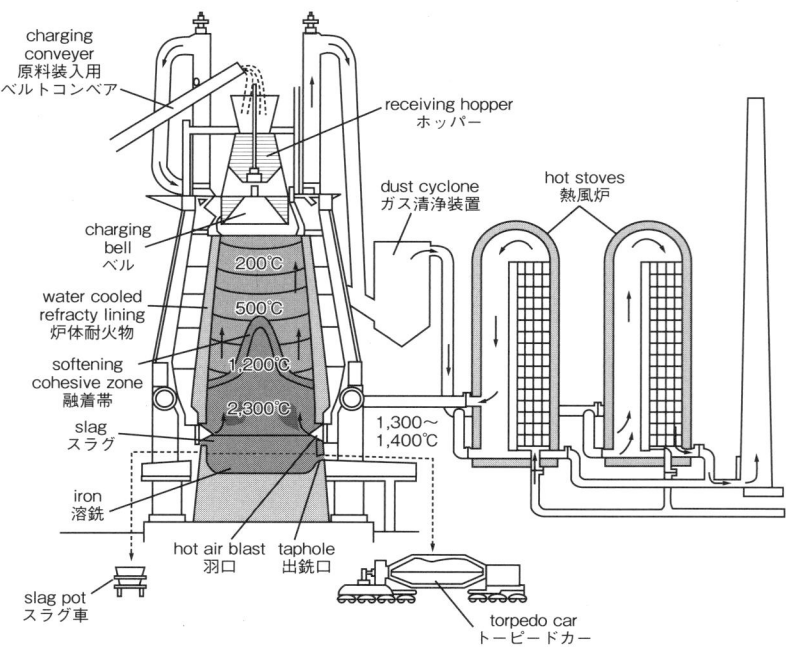

参考図 1-1 製鉄プロセスにおける高炉設備の概要[5]

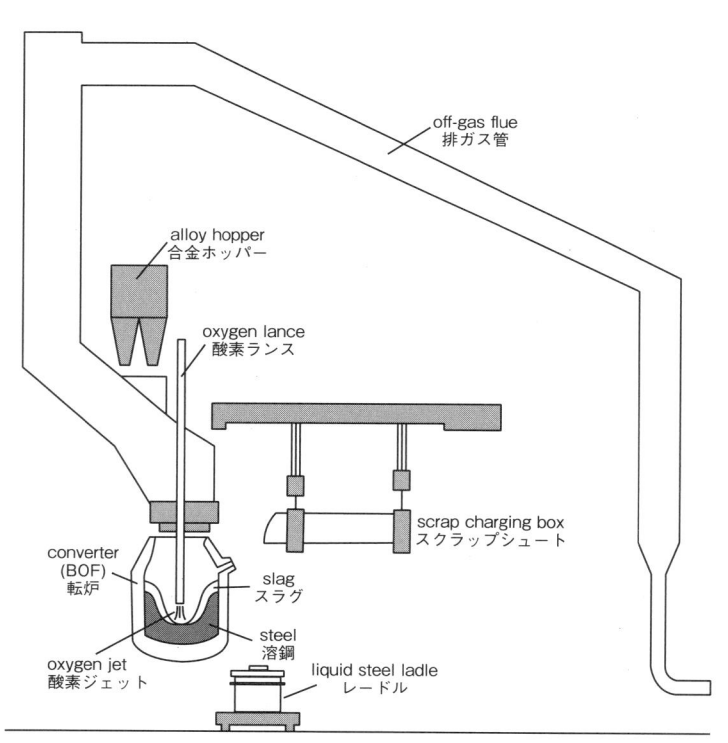

参考図 1-2 製鉄プロセスにおける転炉設備の概要

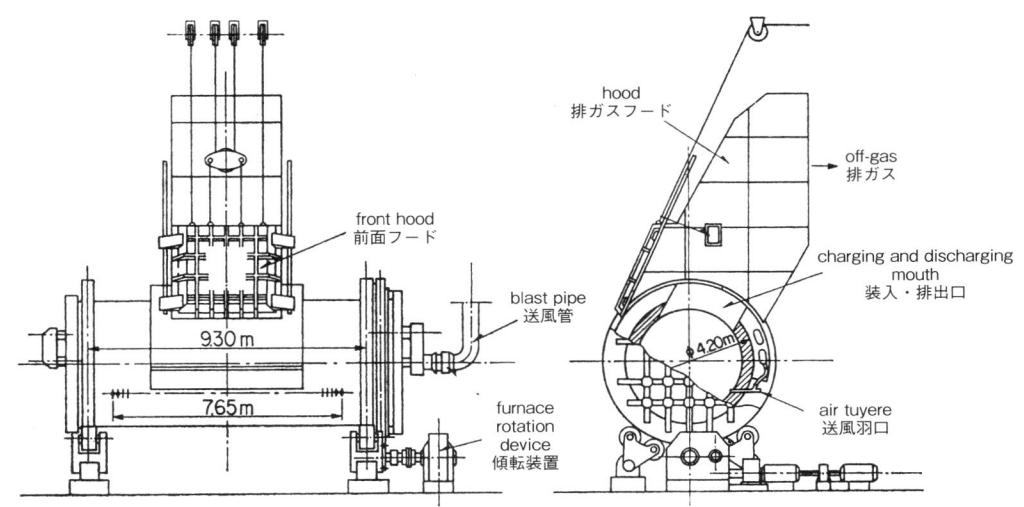

参考図 1-3　Peirce-Smith 転炉設備の概要[6]

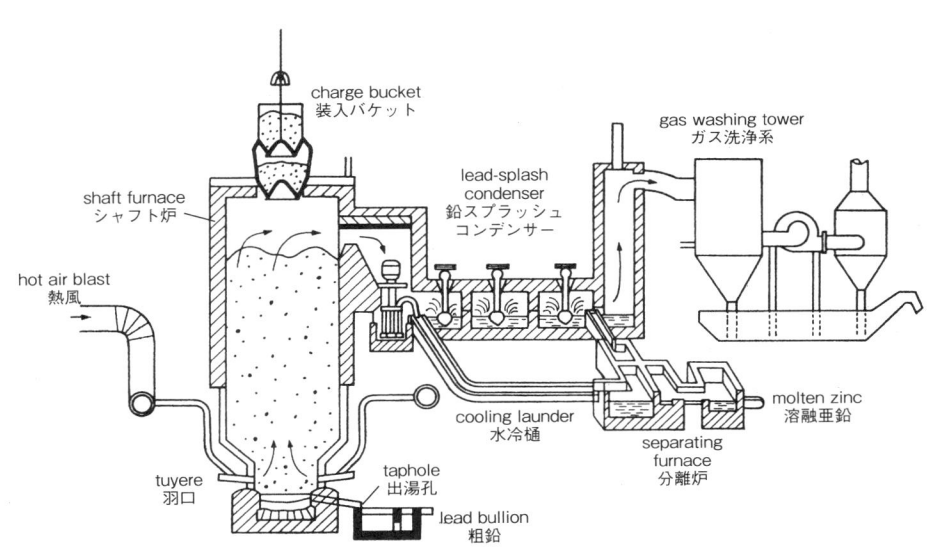

参考図 1-4　Imperial Smelting 炉設備の概要[6]

1. Examples of metallurgical process

マスバランスを取る空間は、ケースにより異なるが (a) マクロに取り扱う場合は全システム、その他 (b) 一次元、(c) 二次元、(d) 三次元を決定する。

特定空間内の成分 i に関する一般的なマスバランスの式は (1-8) 式のようになる。

```
(i のインプット速度) －
        (i のアウトプット速度)
＋(i の生成速度) －(i の消費速度)
＝(i の蓄積速度)
```

"流束"というターミノロジーはラテン語の"流れ"に由来するものであるが、輸送現象論における"流束"は移動する物理量（物質、熱、あるいは運動量）の単位面積当りの移動速度で定義される。

(a) 分子拡散による物質流束（Fick の第一法則）

単位面積当りの物質量（モル量）の移動速度 J は拡散係数 D と濃度勾配（単位長さ当りのモル濃度の変化量）の積で (1-9) 式のように表される。なお、本書では II 章以後、全モル移動速度 $\dot{n}$ (＝J・A) を用いて式を展開する。ここで A は界面積である。

2. Basic equations to describe the metallurgical mass transfer

2.1 Mass balance equation

Law of conservation stipulates that mass must be conserved within a control volume if no matter crosses its boundary.

Types of the control volume are,
(a) whole system
(b) one-dimensional sectional element
 e.g. mass transfer in a pipe
(c) two-dimensional area element
(d) three-dimensional volume element

Mass balance for solute i in a control volume is,
(Rate of input of i) － (Rate of output of i)
＋ (Rate of generation of i) － (Rate of consumption of i)
＝ (Rate of accumulation of i)　　　　　　(1-8)

2.2 Molar flux equation

The word "flux" comes from "fluxus (Latin)" means "flow". In transport phenomena (mass transfer, heat transfer, and momentum transfer), flux is defined as the rate of flow of a property per unit area, which has the [quantity] [time]$^{-1}$ [area]$^{-1}$.

(a) Mass flux (Fick's 1st law) of diffusion

$$J = -D\frac{dC}{dx} \quad (1-9)$$

the rate of mass flow across a unit area (mole/s/m^2)

diffusivity (m^2/s)

concentration gradient (mole/m^3/m)

(b) Heat flux (Fourier's law) of conduction

$$q = -\lambda \frac{dT}{dy} = -\left(\frac{\lambda}{\rho C_p}\right)\frac{d(\rho C_p T)}{dy} \quad (1\text{-}10)$$

the rate of heat flow across a unit area (J/s/m²)
thermal diffusivity (m²/s)
gradient of (thermal energy/volume) (J/m⁴)

(c) Momentum flux (Newton's law) of viscosity

$$\tau_{xy} = -\mu \frac{du_x}{dy} = -\left(\frac{\mu}{\rho}\right)\frac{d(\rho u_x)}{dy} \quad (1\text{-}11)$$

the rate of momentum transfer across a unit area (Ns/s·m²) or shear stress (N/m²)
kinetic viscosity (m²/s)
gradient of (momentum/volume) (Ns/m³/m)

$$(1\text{-}12)$$

以下、参考までに、

(b) 熱流束（Fourierの法則）
(1-10) 式に示すように単位面積当りの熱の移動速度である熱流束（q）は温度勾配と熱伝導率との積で表されるが、熱拡散係数（$\lambda/\rho C_p$）と単位体積当りの熱エネルギー（$\rho C_p T$）の勾配との積とも表現される。ここでρは密度、C_pは比熱である。

(c) 運動量流束（Newtonの法則）
(1-11) 式に示すようにx方向の流体の速度u_xのy方向の勾配に流体の粘度μを掛けた「y軸に垂直な面上のx方向に働くせん断応力」はx方向の単位体積当りの運動量が単位時間、単位面積当りにy方向に移動する量とみなす事ができる。

2.3 Chemical reaction related to mass trasfer

Fig. 1-5 shows the schematic model of the Fe_2O_3 reduction by CO gas which comprises of the following steps.

図1-5に示すようにCOガスによるFe_2O_3（ヘマタイト）の還元は次のステップで進行する。
1) 反応サイトである界面への反応物質（COガス）の移動
2) 界面での化学反応
3) 反応界面からの生成物質（CO_2ガス）の移動

実際の鉄鉱石（Fe_2O_3）の還元反応は（$Fe_2O_3 \rightarrow Fe_3O_4 \rightarrow FeO \rightarrow Fe$）のように段階的に起こるが、ここでは簡略化して(1-13)式のように表現する。

図1-5において界面とはFe₂O₃固体とガスの境界、バルクとはCOガス濃度が一定であるガス本体を指す。

d_i：界面の位置
d_b：バルクの位置
CO_i：界面でのCOガス濃度
CO_{2i}：界面でのCO₂ガス濃度

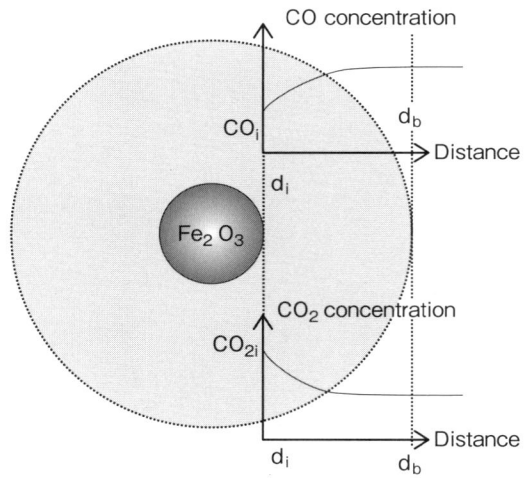

Fig. 1-5 Mass transfer and reaction near the interface between iron oxide and gas.

反応全体の速度は1)～3)の中で最も遅いステップ（律速段階）によって決まる。

1) Transport of reactants to the interface where the reaction occurs.
2) Chemical reaction at the interface.

$$Fe_2O_3 + 3CO = 2Fe + 3CO_2 \quad (1\text{-}13)$$

3) Transport of products away from the interface.

The rate of overall reduction depends on which is the slowest step in the process.

化学反応速度は、一般に反応速度定数と反応にかかわる反応種の濃度の関数で表される。

Rate of chemical reaction = $k_r \cdot f$ (concentration of species), where k_r is a reaction coefficient expressed as

$$k_r = k_o \exp\left(-\frac{E_a}{RT}\right) \quad (1\text{-}14)$$

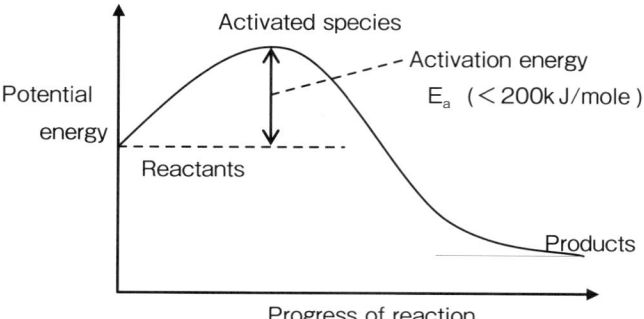

Fig. 1-6 Potential energy profile for an exothermic reaction.[7]

Most metallurgical reactions take place at sufficiently high temperature to make their reaction rate fast or rapid. And apparent activation energy is almost 200 kJ/mole.

ここで取り扱う反応のほとんどは高温反応であるため、反応の活性化エネルギーは〜200 kJ/moleと、物質移動の活性化エネルギー数10 kJ/moleに比べて非常に大きい。

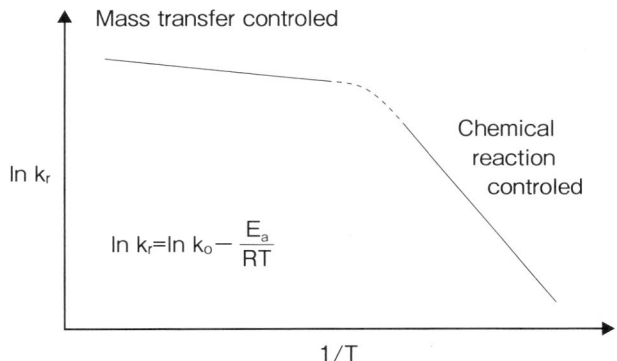

Fig. 1-7　Relationship between (1/T) and ln kr.

It has been found that the productivity of a process is therefore rarely limited by the rate of chemical reaction itself and the transport steps become more important.

プロセスの生産性は反応速度で律速される場合はきわめてまれで、ほとんどが物質移動律速である。そのため、「輸送現象／物質移動」の解釈に基づく制御の提案が非常に重要である。

Two important conceptions are :

1) Equilibrium exists at the interface (because reaction is rapid), so thermodynamics may be used to relate concentration of reactants and products at the interface.

$$A+B=C+D \tag{1-15}$$

$$K=\frac{C_C^* C_D^*}{C_A^* C_B^*} \tag{1-16}$$

2) Rates of solute transfer to and from a reaction site are controlled by transport process and therefore the overall rates of metallurgical reactions are mass transfer limited.

以上のように、高温プロセスの総括反応速度に関する考え方は、

1) 界面での反応は平衡状態であるとみなすことが出来る。
　たとえば (1-15) 式で示す反応が界面で進行する場合、界面では (1-16) 式で示す平衡関係が成り立っていると考える。
　すなわち、界面での熱力学は、界面で反応する種および濃度と、生成する種および濃度を関連付ける役割を果たすことになる。

2) 図1-8に示すような反応界面、および境界層での濃度プロフィールを考える場合、反応サイトへの反応種の移動と、反応サイトからの生成種の移動のどちらか、あるいは両方が、総括反応速度を律速することになる。

$C_{A(b)}$：種Aのバルク濃度
$C_{B(b)}$：種Bのバルク濃度
$C_{C(b)}$：種Cのバルク濃度
$C_{D(b)}$：種Dのバルク濃度
C_A^*：種Aの界面濃度
C_B^*：種Bの界面濃度
C_C^*：種Cの界面濃度
C_D^*：種Dの界面濃度
d：界面からの距離
δ：境界層厚み

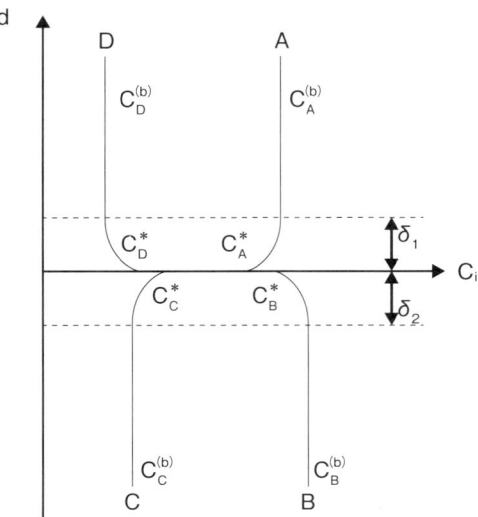

Fig. 1-8 Mass transfer in the boundary layer and reaction at the interface.

次に（1-17）式で示される反応を考える。ここで系の体積は一定であり、ある瞬間の系中の溶質成分Jの濃度を[J]で表す。反応物質R（(1-17)式ではE、Fに相当）の消費速度は$-d[R]/dt$、生成物質P（(1-17)式ではG、Hに相当）の生成底度は$d[P]/dt$となる。

化学量論性を考慮した場合、(1-17)式の各溶質成分の反応速度には(1-18)式の関係が与えられる。

さらに、この化学量論性は、反応における反応量と生成量が等しいという質量保存則のベースでもある。

Then consider a reaction of the form,

$$E + nF \rightarrow mG + H \quad (1\text{-}17)$$

in which at some instant the molar concentration of a participant J is [J] and the volume of the system is costant. The instantaneous rate of consumption of one of reactants at a given time is $-d[R]/dt$, where R is E or F. The rate of formation of one of the products (G or H, which is denoted as P) is $d[P]/dt$. It follows from the stoichiometry for the reaction represented by eq. (1-17) that

$$-\frac{d[E]}{dt} = \frac{1}{m}\frac{d[G]}{dt} = \frac{d[H]}{dt} = -\frac{1}{n}\frac{d[F]}{dt} \quad (1\text{-}18)$$

so there are several rates connected with the reaction. Note that stoichiometry is the relative quantities of reactants and products in the chemical reactions, and is founded on the law of conservation of mass where the total mass of reactants equals the total mass of the products.

II. Steady State Diffusion (Fick's 1st law) and Relating Mass Transfer

1. Molecular diffusion and Fick's 1st law including the effect of convection

Mass transfe may take place by
1) Molecular diffusion :
   Solute transfers from a region of high concentration to low concentration by molecular diffusion.
2) Convection (natural or forced):
   ① Mass transfer by agitation due to stirring (forced convection)
   ② Mass transfer due to densily differences arising from concentration gradient

In most metallurgical systems, both processes occur simultaneously.

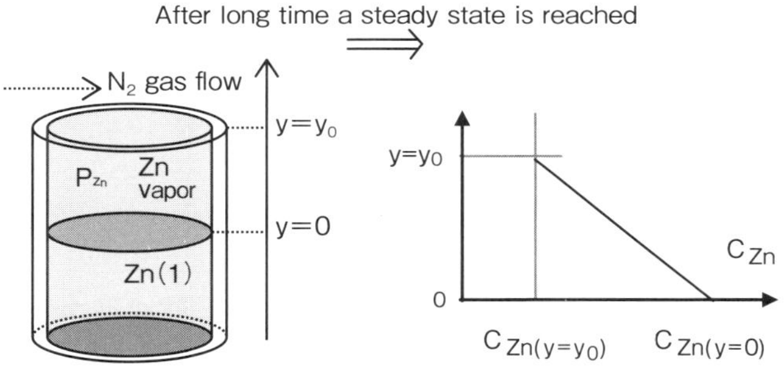

Fig. 2-1 Diffusion of Zn vapor in the vessel.

The following equation descrives the flux of Zn out of the vessel shown in Fig.2-1.

$$J_{Zn} = -D_{Zn}\left[\frac{C_{Zn(y_0)} - C_{Zn(0)}}{y_0}\right] \quad (2\text{-}1)$$

J_{Zn} : mole flux (mole/m^2·s)
C_{Zn} : molar concentration (mole/m^3)
y_0 : distance (m)
D_{Zn} : diffusion coefficient (m^2/sec)

物質移動は次の2つの手段で進行する。

1) 分子拡散
   溶質は高濃度領域から低濃度領域に向かって分子拡散によって移動する。
2) 自然対流もしくは強制対流
   ①攪拌による物質移動(強制対流)
   ②密度差による物質移動(自然対流)

ほとんどの場合、1)、2)、が同時に進行する事が多い。

19世紀半ば、Fickは食塩の水への溶解の実験結果をベースに、すでに発表されていた熱伝導のFourierの法則と同様な数学的法則が成り立つと仮定し、(2-2)式で示されるような、モル流束J_{Zn}を定義した。これがFickの第一法則である。
たとえば図2-1の左図のように、液体亜鉛の上部($y=0 \sim y_0$)が亜鉛蒸気で満たされている空間を考える。
なお、P_{Zn}($y=0$)とC_{Zn}($y=0$)、P_{Zn}($y=y_0$)とC_{Zn}($y=y_0$)には以下の関係がある。

$$P_{Zn(y=0)} = \frac{n_{Zn(y=0)}}{V}RT$$
$$= C_{Zn(y=0)}RT$$

$$P_{Zn(y=y_0)} = \frac{n_{Zn(y=y_0)}}{V}RT$$
$$= C_{Zn(y=y_0)}RT$$

容器の上部は蓋で覆われており、容器外部は十分な量の窒素ガスが流れている。ある瞬間にこの容器の蓋を取り除くことによって、液体亜鉛上部の亜鉛ガスに濃度勾配が生じるが、十分な時間が経過した後、この空間を通して移動する亜鉛の

モル流束 J_{Zn}(後述する $\frac{\dot{n}_{Zn}}{A}$ と同等) は(2-1)式で表される。

全モル濃度を C_T、モル分率を X_{Zn} とすると(2-4)式のようになる。

ここでモル流束と濃度勾配の比例定数 D_i は拡散係数と呼ばれ、拡散する物質と媒体となる物質の組み合わせで決まる。表1-1にガスや溶融金属、固体金属中の拡散係数のオーダーを示す。

$$J_{Zn} = -D_{Zn}\frac{dC_{Zn}}{dy} \qquad (2\text{-}2)$$

Fick's 1st law of eq. (2-2) expresses molecular diffusion rate.

D_{Zn} is not a function of C_{Zn}, which is generally true for gases and dilute solutions.

If we can specify that

$$C_{Zn} = \underbrace{C_T}_{\text{molar density}} \cdot \underbrace{X_{Zn}}_{\text{molar fraction}} \qquad (2\text{-}3)$$

$$\therefore J_{Zn} = -D_{Zn}C_T\frac{dX_{Zn}}{dy} \qquad (2\text{-}4)$$

Table. 1-1 Typical values of diffusion coefficient

a)	gases	$1\text{cm}^2/\text{s}$
b)	liquids	$D_{c,Fe} = 5\times10^{-4}\text{ cm}^2/\text{s}$ at 1400℃
c)	solids	$10^{-8} \sim 10^{-9}\text{ cm}^2/\text{s}$

次に、分子拡散に及ぼす対流の影響についてKirkendall効果を例に説明する。

図2-2に示すような金—ニッケル間の拡散を考える。金の拡散速度の方がニッケルより速いため、両金属間の境界に挿入したMoマーカーは、所定の時間経過後、金寄りに移動したかのように見える(Kirkendall効果)。このように各成分の拡散速度の差によって生じる拡散量から求められる拡散係数は相対的なものであり、相互拡散係数(interdiffusion coefficient)と呼ばれる。なお、Kirkendall効果に関しては、厳密には原子空孔の挙動を考慮する必要がある。

Next, as an example of the effect of convection on the molecular diffusion, "Kirkendall effect" is shown in Fig 2-2.

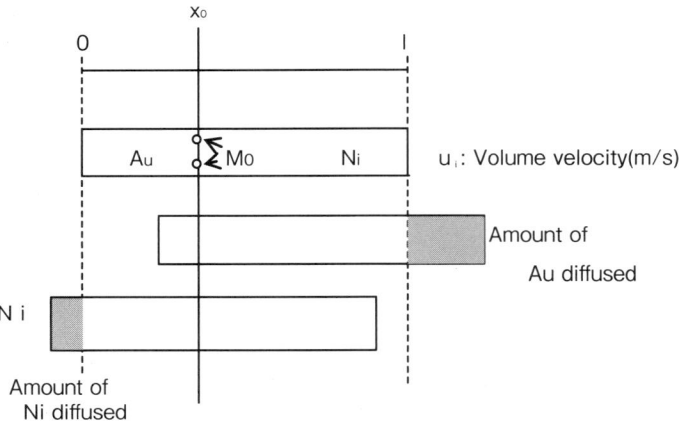

Fig. 2-2 Kirkendall effect.

By Fick's 1st law

$$J_{Au} = -D_{Au,Ni}\frac{\partial C_{Au}}{\partial x} \qquad (2\text{-}5)$$

Flux due to movement of bar as a result of the difference in relative diffusion of Au and Ni,

$$\frac{\dot{n}_{Au}}{A} = \underbrace{u_m}_{\text{volume velocity}} \cdot C_{Au} \qquad (2\text{-}6)$$

マーカーを通過する流束は、分子拡散と相互拡散の差によって生じる体積流れの和で表すことができる（(2-7)式）。

Total flux relative to fixed frame of reference such as innert marker,

$$\frac{\dot{n}_{Au}}{A} = -D_{Au,Ni}\frac{\partial C_{Au}}{\partial x} + u_m \cdot C_{Au} \qquad (2\text{-}7)$$

ここで体積流れ u_m は多元系の場合(2-9)式のように表され、2元系の場合は(2-12)式のようになる。

In multi-component system, total flux is

$$\frac{\dot{n}_i}{A} = \underline{u_i} \cdot \underline{C_i} \qquad (2\text{-}8)$$

mole/cm²·s   cm/s   mole/cm³

where u_i is diffusion velocity of component i.

Mole center velocity (opp. is mole local velocity) u_m of a fluid is given by

$$u_m = \frac{u_1 C_1 + u_2 C_2 + \cdots\cdots + u_n C_n}{C_1 + C_2 + \cdots\cdots + C_n} \qquad (2\text{-}9)$$

$$= \frac{\frac{\dot{n}_1}{A} + \frac{\dot{n}_2}{A} + \cdots\cdots + \frac{\dot{n}_n}{A}}{\sum_{i=1}^{n} C_i} \qquad (2\text{-}10)$$

$$= \frac{\frac{\dot{n}_1}{A} + \frac{\dot{n}_2}{A} + \cdots\cdots + \frac{\dot{n}_n}{A}}{C_T} \qquad (2\text{-}11)$$

In a binary system

$$u_m = \frac{u_1 C_1 + u_2 C_2}{C_1 + C_2} \qquad (2\text{-}12)$$

Fick's 1st law / Mass transfer in a binary system (consists of two components ⟨1⟩ and ⟨2⟩) is expressed,

成分⟨1⟩、⟨2⟩から成る2元系の、流れを考慮した定常拡散による物質移動は(2-14) 式のように表される。

$$\frac{\dot{n}_1}{A} = -D_{1,2}\frac{\partial C_1}{\partial x} + C_1 \cdot u_m \qquad (2\text{-}13)$$

$$= -D_{1,2}\frac{\partial C_1}{\partial x} + C_1 \boxed{\frac{u_1 C_1 + u_2 C_2}{C_1 + C_2}} \text{ mole center velocity} \qquad (2\text{-}14)$$

1. Molecular diffusion (Fick's 1st law) including the effect of convection

$$= -D_{1,2}\frac{\partial C_1}{\partial x} + C_1\frac{(\frac{\dot{n}_1}{A}+\frac{\dot{n}_2}{A})}{C_1+C_2} \qquad (2\text{-}15)$$

$$= -D_{1,2}\frac{\partial C_1}{\partial x} + (\frac{C_1}{C_1+C_2})(\frac{\dot{n}_1}{A}+\frac{\dot{n}_2}{A}) \qquad (2\text{-}16)$$

$$= \underline{-D_{1,2}\frac{\partial C_1}{\partial x}} + \underline{X_1(\frac{\dot{n}_1}{A}+\frac{\dot{n}_2}{A})} \qquad (2\text{-}17)$$

due to molecular diffusion

due to bulk motion of medium as a result of either convection and/or diffusion

X_1 : mole fraction of component $\langle 1 \rangle$.

$D_{1,2}$: diffusion coefficient of species $\langle 1 \rangle$ in $\langle 2 \rangle$.

<Note>

For dilute solusion $X_1 \ll 1$,

$$\frac{\dot{n}_1}{A} \simeq -D_{1,2}\frac{\partial C_1}{\partial x} \qquad (2\text{-}18)$$

In a binary mixture,

$$\frac{\dot{n}_1}{A} = -D_{1,2}\frac{\partial C_1}{\partial y} + X_1(\frac{\dot{n}_1}{A}+\frac{\dot{n}_2}{A}) \qquad (2\text{-}19)$$

$$\frac{\dot{n}_2}{A} = -D_{2,1}\frac{\partial C_2}{\partial y} + X_1(\frac{\dot{n}_1}{A}+\frac{\dot{n}_2}{A}) \qquad (2\text{-}20)$$

$$\therefore \frac{\dot{n}_1}{A}+\frac{\dot{n}_2}{A} = -D_{2,1}\frac{\partial C_2}{\partial y} - D_{1,2}\frac{\partial C_1}{\partial y}$$

$$+ (X_1+X_2)(\frac{\dot{n}_1}{A}+\frac{\dot{n}_2}{A}) \qquad (2\text{-}21)$$

$$\therefore D_{2,1}\frac{\partial C_2}{\partial y} + D_{1,2}\frac{\partial C_1}{\partial y} = 0 \qquad (2\text{-}22)$$

On the other hand

$$C_1+C_2=C_T \text{ (const.)} \qquad (2\text{-}23)$$

$$\therefore \frac{\partial C_1}{\partial y}+\frac{\partial C_2}{\partial y}=0 \qquad (2\text{-}24)$$

From eqs. (2-22) and (2-24)

$$D_{1,2}=D_{2,1} \qquad (2\text{-}25)$$

Thus in a binary system there is only one inter-diffusion coefficient.

2. Steady state diffusion problems
2.1 Diffusion of component ⟨1⟩ in stagnant medium ⟨2⟩

最初に、定常拡散の事例として、滞留している媒体⟨2⟩中の成分⟨1⟩の拡散について考える。

図2-3に示すような液体燃料粒子が周囲に蒸発する場合の気体燃料の濃度分布を求める。

液体粒子の中心からrの距離にある厚さdrの薄膜を考える（粒子の径をr_0とし、薄膜外側までの径をr_∞とする）。

Consider a fuel droplet injected into a furnace and we are interested in determining the rate of evaporation of the droplet and the concentration profiles of the fuel vapor in the surrounding medium.

Determine the concentration profile of the fuel vapor and mass flux, based on the assumption of steady state and uniform vaporization in all direction.

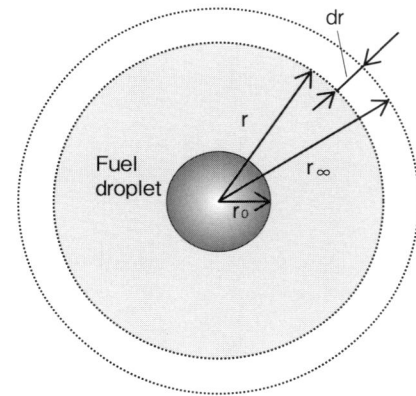

Fig. 2-3 Evaporation of the fuel droplet.

⟨Equations⟩

① mass balance

$$\text{Rate of input} = \frac{\dot{n}_1}{A} \cdot A \tag{2-26}$$

$$\text{Rate of output} = \frac{\dot{n}_1}{A} \cdot A + \frac{d}{dr}\left(\frac{\dot{n}_1}{A} \cdot A\right) dr \tag{2-27}$$

Rate of generation, consumption = 0 (2-28)
(no reaction)

Rate of accumulation = 0 (2-29)
(steady state)

この薄膜における物質収支は以下のようになる。

被膜の内側から流入する成分⟨1⟩のモル流束は$\frac{\dot{n}_1}{A}$であるため流入の全モル流束は(2-26)式のように表される。一方、薄膜から流出する成分⟨1⟩は(2-27)式のように表される。この薄膜内での成分⟨1⟩の生成や蓄積は無いと考えると、全体のマスバランスは、(2-30)式、最終的には(2-31)式のように表される。

$$\therefore \overbrace{\frac{\dot{n}_1}{A}A}^{\text{input}} = \overbrace{\frac{\dot{n}_1}{A}A + \frac{d}{dr}\left(\frac{\dot{n}_1}{A} \cdot A\right)dr}^{\text{output}} \tag{2-30}$$

⟨参考⟩

薄膜内の流束勾配$\frac{d}{dr}\left(\frac{\dot{n}_1}{A}A\right) = \frac{d\dot{n}_1}{dr}$に薄膜厚みdrを乗じたものが出口側の濃度流

Finally,
$$\frac{d}{dr}\left\{\left(\frac{\dot{n}_1}{A}\right)A\right\}=0 \qquad \therefore \frac{d}{dr}(\dot{n}_1)=0 \quad (2\text{-}31)$$

② flux equation

$$\frac{\dot{n}_1}{A} = -D_{1,2}\frac{dC_1}{dr} + X_1\left(\frac{\dot{n}_1}{A}+\frac{\dot{n}_2}{A}\right) \quad (2\text{-}32)$$
$$\underset{0 \text{ for stagnant medium ②}}{}$$

$$\therefore \frac{\dot{n}_1}{A} = -\frac{D_{1,2}}{(1-X_1)}\frac{dC_1}{dr} \quad (2\text{-}33)$$

$$\therefore \dot{n}_1 = -\frac{D_{1,2}\cdot C_T}{(1-X_1)}\cdot 4\pi r^2 \frac{dX_1}{dr} \quad (2\text{-}34)$$

$(\because A=4\pi r^2, \ C_1=C_T X_1)$

$$\therefore \frac{d}{dr}\left\{\underline{\frac{D_{1,2}C_T}{(1-X_1)}4\pi r^2 \frac{dX_1}{dr}}\right\}=0 \quad (2\text{-}35)$$
$$\text{const.}$$

$$\frac{r^2}{1-X_1}\frac{dX_1}{dr}=\alpha(\text{const.}) \quad (2\text{-}36)$$

$$\therefore \ln(1-X_1)=\frac{\alpha}{r}+\beta \quad (2\text{-}37)$$

B. C.
$r=r_0, \quad X_1=X_{1,0} \quad (2\text{-}38)$
$r=r_\infty, \quad X_1=X_{1,\infty} \quad (2\text{-}39)$

$$\therefore \ln(1-X_{1,0})=\frac{\alpha}{r_0}+\beta \quad (2\text{-}40)$$

$$\ln(1-X_{1,\infty})=\frac{\alpha}{r_\infty}+\beta \quad (2\text{-}41)$$

$(2\text{-}40)-(2\text{-}41) \quad \ln\frac{1-X_{1,0}}{1-X_{1,\infty}}=\alpha\left(\frac{1}{r_0}-\frac{1}{r_\infty}\right) \quad (2\text{-}42)$

$$\therefore \alpha=\left(\frac{1}{r_0}-\frac{1}{r_\infty}\right)^{-1}\ln\frac{1-X_{1,0}}{1-X_{1,\infty}} \quad (2\text{-}43)$$

$$\therefore \beta=\ln(1-X_{1,0})-\frac{1}{r_0}\left(\frac{1}{r_0}-\frac{1}{r_\infty}\right)^{-1}\ln\frac{1-X_{1,0}}{1-X_{1,\infty}}$$
$$(2\text{-}44)$$

$$\therefore \ln(1-X_1)=\frac{1}{r}\left(\frac{1}{r_0}-\frac{1}{r_\infty}\right)^{-1}\ln\frac{1-X_{1,0}}{1-X_{1,\infty}}+$$

出速度の追加分となる。

(2-32)式において $\frac{\dot{n}_2}{A}=0$ （成分〈2〉は滞留しているので）とおくと(2-33)式となり、(2-34)式、(2-35)式が得られる。
$\frac{d}{dr}(\dot{n}_1)=0$ であるため(2-35)式のように書き直される。

$\frac{r^2}{1-X_1}\frac{dX_1}{dr}$ を定数 (α) とおいて (2-37) 式を得る。β は積分定数である。

ここで $r=r_0$（被膜の内側）で成分〈1〉のモル濃度 $X_1=X_{1,0}$、$r=r_\infty$（被膜の外側）で成分〈1〉のモル濃度 $X_1=X_{1,\infty}$ とすると、(2-40)式、(2-41)式が得られる。

(2-40)式−(2-41)式でβを消去すると、(2-42)式となり、これよりαとβが求まる。

以上のようにして、被膜中の成分の濃度プロフィールは(2-47)式のように表すことができる。

$$\ln(1-X_{1,0}) - \frac{1}{r_0}\left(\frac{1}{r_0} - \frac{1}{r_\infty}\right)^{-1} \ln\frac{1-X_{1,0}}{1-X_{1,\infty}} \quad (2\text{-}45)$$

$$\therefore \ln\frac{1-X_1}{1-X_{1,0}} = \left(\frac{1}{r} - \frac{1}{r_0}\right)\left(\frac{1}{r_0} - \frac{1}{r_\infty}\right)^{-1} \ln\frac{1-X_{1,0}}{1-X_{1,\infty}} \quad (2\text{-}46)$$

$$\therefore \underbrace{\left(\frac{1}{r_0} - \frac{1}{r_\infty}\right)\ln\left(\frac{1-X_1}{1-X_{1,0}}\right) = \left(\frac{1}{r_0} - \frac{1}{r}\right)\ln\left(\frac{1-X_{1,\infty}}{1-X_{1,0}}\right)}_{\text{concentration gradient}} \quad (2\text{-}47)$$

(2-47)式の両辺を、X_1 および r で微分すると、(2-48)式のようになる。したがって dX_1/dr は (2-49) 式のように表される。

differentiating eq. (2-47) by X_1 & r

$$\left(\frac{1}{r_0} - \frac{1}{r_\infty}\right)\left(\frac{-1}{1-X_{1,0}}\right)dX_1 = \frac{1}{r^2}\ln\left(\frac{1-X_{1,\infty}}{1-X_{1,0}}\right)dr \quad (2\text{-}48)$$

$$\therefore \frac{dX_1}{dr} = \frac{1}{r^2} \cdot \ln\left(\frac{1-X_{1,\infty}}{1-X_{1,0}}\right)\frac{-(1-X_{1,0})}{\left(\frac{1}{r_0} - \frac{1}{r_\infty}\right)} \quad (2\text{-}49)$$

$r = r_0$ における $\frac{dX_1}{dr}$ は (2-51) 式のようになるため $\dot{n}_{1(r=r_0)}$ は (2-52) 式のようになる。最終的に成分1のモルフラックス式は (2-53) 式のようになる。

$$\therefore \left(\frac{dX_1}{dr}\right)_{r=r_0} = \frac{-(1-X_{1,0})}{r_0^2\left(\frac{1}{r_0} - \frac{1}{r_\infty}\right)}\ln\left(\frac{1-X_{1,\infty}}{1-X_{1,0}}\right) \quad (2\text{-}50)$$

$$= -\frac{(1-X_{1,0})}{(r_\infty - r_0)} \cdot \frac{r_\infty}{r_0}\ln\left(\frac{1-X_{1,\infty}}{1-X_{1,0}}\right) \quad (2\text{-}51)$$

$$\dot{n}_{1(r=r_0)} = -\frac{D_{1,2}C_T 4\pi r_0^2}{1-X_{1,0}}\left(\frac{dX_1}{dr}\right)_{r=r_0} \quad (2\text{-}52)$$

$$\therefore \frac{\dot{n}_{1(r=r_0)}}{4\pi r_0^2} = \frac{D_{1,2}C_T \cdot r_\infty}{r_0(r_\infty - r_0)}\ln\left(\frac{1-X_{1,\infty}}{1-X_{1,0}}\right) \quad (2\text{-}53)$$

The equation can be further rearranged to be a more suitable form;

より簡潔な表現を取るならば、$X_{2,\infty} = 1 - X_{1,\infty}$, $X_{2,0} = 1 - X_{1,0}$ であることから、(2-54)式のようになり、(2-57)式で表される「対数平均」$X_{2,\text{ln}}$ を用いると (2-56) 式が得られる。

$$\frac{\dot{n}_1}{4\pi r_0^2} = \frac{D_{1,2}C_T r_\infty}{r_0(r_\infty - r_0)}\ln\frac{X_{2,\infty}}{X_{2,0}} \quad (2\text{-}54)$$

$$= \frac{D_{1,2}C_T r_\infty}{r_0(r_\infty - r_0)}\left(\ln\frac{X_{2,\infty}}{X_{2,0}}\right)\frac{X_{2,\infty} - X_{2,0}}{X_{2,\infty} - X_{2,0}} \quad (2\text{-}55)$$

$$= \underbrace{\frac{D_{1,2}C_T r_\infty}{r_0(r_\infty - r_0)}\frac{1}{X_{2,\text{ln}}}}_{\text{conductance}} \underbrace{(X_{2,\infty} - X_{2,0})}_{\text{driving force}} \quad (2\text{-}56)$$

$$\text{log mean mole fraction} \quad X_{2,ln} = \frac{X_{2,\infty} - X_{2,0}}{\ln \frac{X_{2,\infty}}{X_{2,0}}} \quad (2\text{-}57)$$

Simplification;

1) $r_\infty - r_0 \to \delta \ll r_\infty,\ r_0,\quad \frac{r_\infty}{r_0} \simeq 1$ \quad (2-58)

2) X_1 is small \quad ($X_{2,\infty} - X_{2,0}$ is small) \quad (2-59)

by L'Hopital's rule

$$X_{2,ln} = \frac{X_{2,\infty} - X_{2,0}}{\ln \frac{X_{2,\infty}}{X_{2,0}}} \simeq 1 \quad (2\text{-}60)$$

$$\frac{\dot{n}_1}{4\pi r_0^2} = \frac{D_{1,2} C_T}{\delta}(X_{1,0} - X_{1,\infty}) \quad (2\text{-}61)$$

ここで、(2-58)式のようにδを定義し、さらに(2-60)式のようにX_1とX_2を置き換えることによって、(2-61)式のように表すことができる。

2.2 Diffusion accompanied by a fast heterogeneous chemical reaction

(1) Equimolar counter diffusion

$$Fe_2O_3 + 3CO \to 2Fe + 3CO_2 \quad (2\text{-}62)$$

equimolar counter diffusion

続いて、高速の不均一反応を伴う拡散の事例について述べる。

まず、(2-62)式で表される、等モル相互拡散の場合を考える。
(この反応は、前述したように$3Fe_2O_3 + CO = 2Fe_3O_4 + CO_2$、$Fe_3O_4 + CO = 3FeO + CO_2$、$FeO + CO = Fe + CO_2$のように段階的に進行するが、ここでは簡略化している)。

図2-4に示すように、固体の反応体Fe_2O_3の前面に厚さδの濃度遷移境界層を仮定する。反応するガスCOと生成するガスCO_2の移動は対抗し、そのモル数は等しいため、この境界層における拡散は「等モル相互拡散」と呼ばれる。

Choose a system as shown in Fig. 2-4.

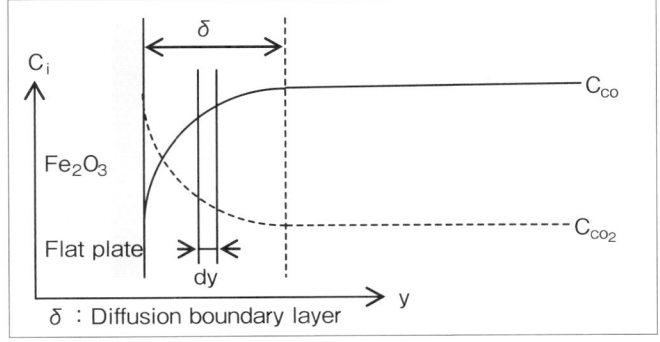

Fig. 2-4 Concentration profile at the interface between Fe_2O_3 and gas.

<Assumptions>

a) Steady state.

b) Product layer Fe formed does not adhere to surface but falls off. Otherwise we have to consider its resistance to diffision of CO and CO_2.

c) Fast chemical reaction, so concentrations of reactants and products at interface are at equilibrium for the reaction.

<Equations>

Assuming the transport of CO is the slowest step in the process, determine $\dot{n}_{co}$ from the following equations.

① stoichiometry

$$\frac{\dot{n}_{co}}{A} = -\frac{\dot{n}_{co_2}}{A} = \frac{3\dot{n}_{Fe_2O_3}}{A} \tag{2-63}$$

(2-63)式から明らかなように、COガスの生成モル流束$\dot{n}_{co}$、CO_2ガスの生成モル流束$\dot{n}_{co_2}$、Fe_2O_3の反応モル流束$\dot{n}_{Fe_2O_3}$は化学量論的関係から(2-63)式で関係づけられる。
Aは反応面積で一定である。

② mass balance in "dy"

$\Bigg($ Rate of input of CO is equivalent to rate of output of CO.

$$-\frac{\dot{n}_{co}}{A}A = -\frac{\dot{n}_{co}}{A}A - \frac{d}{dy}\left(\frac{\dot{n}_{co}}{A}\right)A\,dy \tag{2-64}$$

(A is constant and y>0)

$$\frac{d}{dy}\left(\frac{\dot{n}_{co}}{A}\cdot A\right)dy = 0 \tag{2-65}$$

$$\frac{d}{dy}\left(\frac{\dot{n}_{co}}{A}\right) = 0 \tag{2-66}$$

続いて、境界層の厚みdyの微小領域での物質収支の式、流束の式はそれぞれ(2-66)、(2-68)式のようになる。

③ flux equation

$$\frac{\dot{n}_{co}}{A} = -D_{co,co_2}\cdot C_T\frac{dX_{co}}{dy} + \left(\frac{\dot{n}_{co}}{A} + \frac{\dot{n}_{co_2}}{A}\right)X_{co} \tag{2-67}$$

Because no mole center velocity, eq. (2-67) becomes eq. (2-68).

$$\frac{\dot{n}_{co}}{A} = -D_{co,co_2}\cdot C_T\frac{dX_{co}}{dy} \tag{2-68}$$

Substitute in the mole balance eq. (2-66)

$$\frac{d}{dy}\left(-D_{co,co_2}\cdot C_T\cdot\frac{dX_{co}}{dy}\right) = 0 \tag{2-69}$$

$$\therefore \frac{dX_{co}}{dy} = \alpha \,(\text{const}) \qquad (2\text{-}70)$$

$$\text{B.C.} \quad y=0, \quad X_{co} = X_{co,0} \qquad (2\text{-}71)$$

$$y=L, \quad X_{co} = X_{co,L} \qquad (2\text{-}72)$$

$$\therefore \boxed{\frac{y}{L} = \frac{X_{co}-X_{co,0}}{X_{co,L}-X_{co,0}}} \rightarrow \text{linear profile} \qquad (2\text{-}73)$$

これらの式から(2-70)式が得られ、境界条件((2-71)、(2-72)式)にしたがって、(2-73)式が得られる。

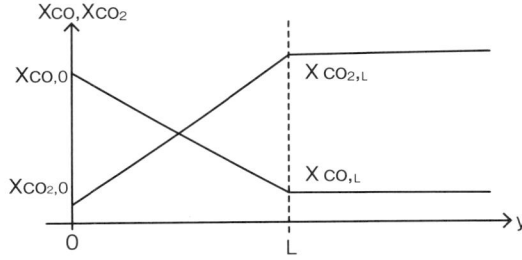

Fig. 2-5 Linear concentration profile in the boundary layer.

from eq. (2-68)

$$\boxed{\frac{\dot{n}_{co}}{A} = -D_{co,co_2} \cdot C_T \frac{X_{co,L}-X_{co,0}}{L}} \qquad (2\text{-}74)$$

Since the overall rate of reduction is diffusion control,

$$\boxed{\frac{\dot{n}_{Fe_2O_3}}{A} = \frac{|\dot{n}_{co}|}{3A} = \frac{1}{3}\frac{D_{co,co_2}C_T}{L}(X_{co,L}-X_{co,0})}$$

$$(2\text{-}75)$$

(2-73)式で示される濃度分布を(2-68)、(2-63)式に代入すると、それぞれCOガスおよびFe_2O_3の生成モル流束、反応モル流束を表す式を得る。

<Note>

In this case rate of reduction calculated is valid for initial stage of reaction. Once a porous product layer is formed, it offers resistance to diffusion of CO and CO_2.

(2) Non-equimolar counter diffusion
(Fast heterogeneous reaction)

続いて、非等モル相互拡散の事例について説明する。

図2-6に示すように固体X, Yと気体G_1, G_2との間に(2-76)式で示される反応が進行している。固体前面のガス拡散境界層（厚みδ）中では反応気体G_1と対抗する生成気体G_2のモル数は同等ではない。

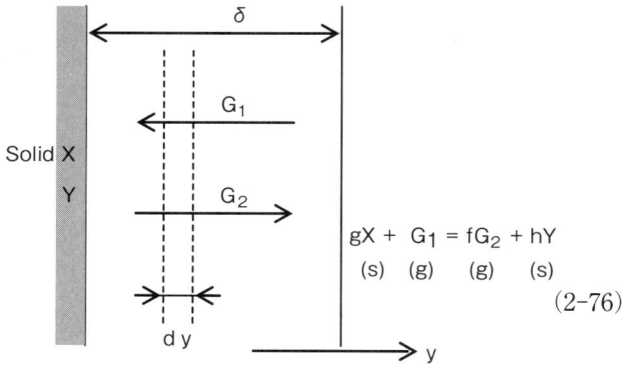

$$gX + G_1 = fG_2 + hY$$
$$(s) \quad (g) \quad (g) \quad (s)$$
(2-76)

Fig. 2-6 Chemical reaction and mass transfer at the interface between solid and gas.

<Assumptions>

a) Steady state.
b) Fast heterogeneous reaction.
c) Reaction control lies in the transport of gaseous reactant to the interface.

Assuming product solid Y does not offer any resistance to diffusion, overall chemical reaction is limitted by rate of transport of G_1 in diffusion layer.

化学量論性から(2-77)式が成り立つ。続いて境界層中の微小厚みdyでのマスバランスは(2-78)式、流束の式は(2-79)式で表される。(2-77)、(2-79)式よりG_1の流束は(2-80)式のようになる。

① stoichiometry

$$\frac{\dot{n}_{G_2}}{A} = -f\frac{\dot{n}_{G_1}}{A} \quad , \quad n_X = g|\dot{n}_{G_1}| \qquad (2\text{-}77)$$

② mass balance in dy

$$\frac{d}{dy}\left(\frac{\dot{n}_{G_1}}{A}\right) = 0 \qquad (2\text{-}78)$$

③ flux equation

$$\frac{\dot{n}_{G_1}}{A} = -\frac{D_{G_1,G_2} \cdot C_T \cdot dX_{G_1}}{dy} + X_{G_1}\left(\frac{\dot{n}_{G_1}}{A} + \frac{\dot{n}_{G_2}}{A}\right) \quad (2\text{-}79)$$

Substituting for $\dfrac{\dot{n}_{G_1}}{A}$ from stoichiometry

$$\frac{\dot{n}_{G_1}}{A} = -D_{G_1,G_2} \cdot C_T \frac{dX_{G_1}}{dy} + (1-f)\frac{\dot{n}_{G_1}}{A} \cdot X_{G_1} \quad (2\text{-}80)$$

Arranging eq. (2-80)

$$\frac{\dot{n}_{G_1}}{A}[1-(1-f)X_{G_1}] = -D_{G_1,G_2} \cdot C_T \frac{dX_{G_1}}{dy} \quad (2\text{-}81)$$

$$\therefore \frac{\dot{n}_{G_1}}{A} = -\frac{D_{G_1,G_2} \cdot C_T}{\{1-(1-f)X_{G_1}\}} \frac{dX_{G_1}}{dy} \quad (2\text{-}82)$$

$$\therefore \frac{d}{dy}\left\{\frac{D_{G_1,G_2} \cdot C_T}{[1-(1-f)X_{G_1}]} \frac{dX_{G_1}}{dy}\right\} = 0 \quad (2\text{-}83)$$

(2-78)式と(2-82)式より(2-83)式が得られ、G_1の反応界面における濃度（反応平衡にあるモル分率 $X_{G_1,0}$）と境界層厚みδにおけるモル分率 $X_{G_1,\delta}$ を境界条件として解くと(2-87)式で示される濃度分布が得られる。

B.C. $y=0$, $X_{G_1} = X_{G_1,0}$ (equilibrium concentration)

$$(2\text{-}84)$$

$y=\delta$, $X_{G_1} = X_{G_1,\delta}$ (bulk concentration)

$$(2\text{-}85)$$

$$\frac{D_{G_1,G_2} \cdot C_T}{[1-(1-f)X_{G_1}]} \frac{dX_{G_1}}{dy} = \alpha \text{ (const.)} \quad (2\text{-}86)$$

from B.C.,

$$\boxed{\ln\left\{\frac{1-(1-f)X_{G_1}}{1-(1-f)X_{G_1,0}}\right\} = \frac{y}{\delta}\ln\left\{\frac{1-(1-f)X_{G_1,\delta}}{1-(1-f)X_{G_1,0}}\right\}} \quad (2\text{-}87)$$

concentration profile

from eq. (2-82)

$$\int_0^\delta \frac{\dot{n}_{G_1}}{A} dy = \int_{X_{G_1,0}}^{X_{G_1,\delta}} -\frac{D_{G_1,G_2} \cdot C_T}{1-(1-f)X_{G_1}} dX_{G_1} \quad (2\text{-}88)$$

Because $\frac{\dot{n}_{G_1}}{A}$ is a constant,

$$\boxed{\frac{\dot{n}_{G_1}}{A} = \frac{D_{G_1,G_2} \cdot C_T}{(1-f)\delta}\ln\left\{\frac{1-(1-f)X_{G_1,\delta}}{1-(1-f)X_{G_1,0}}\right\}} \quad (2\text{-}89)$$

molar flux

これよりG_1のモル流束、Xのモル流束は、それぞれ(2-89)、(2-91)式のように表される。

2. Steady state diffusion problems

$$\boxed{\begin{aligned}\frac{\dot{n}_X}{A}&=g\left|\frac{\dot{n}_{G_1}}{A}\right|\\&=\frac{gD_{G_1,G_2}C_T}{(1-f)\delta}\ln\left\{\frac{1-(1-f)X_{G_1,\delta}}{1-(1-f)X_{G_1,0}}\right\}\end{aligned}} \quad (2\text{-}90)$$

reaction rate for diffusion control

Consider the case "G_2 is stagnant"; f=0, concentration profile is,

$$\frac{1-X_{G_1}}{1-X_{G_1,0}}=\left(\frac{1-X_{G_1,\delta}}{1-X_{G_1,0}}\right)^{\frac{y}{\delta}} \quad (2\text{-}91)$$

similarly,

$$\frac{\dot{n}_{G_1}}{A}=\frac{D_{G_1,G_2}C_T}{\delta}\ln\left(\frac{1-X_{G_1,\delta}}{1-X_{G_1,0}}\right) \quad (2\text{-}92)$$

If f = 1, it is "equimolar counter diffusion". Expression becomes undetermined. We have to derive the equations from 1st principles. See 2.2(1).

ここで f=0 の場合（G_2 が淀んでいる場合）は2.1のケースに、f=1 の場合は2.2(1)の等モル拡散のモル流束の式に相当することになる。

2. 3 Diffusion accompanied by a homogeneous chemical reaction

均一化学反応を伴う拡散の事例を紹介する。

図2-7に示すように、液体亜鉛の入った容器の上部を窒素と微量の酸素を含むガスが多量に流れている。

酸素は容器中に侵入して亜鉛蒸気と反応して酸化亜鉛のヒュームを生成する。

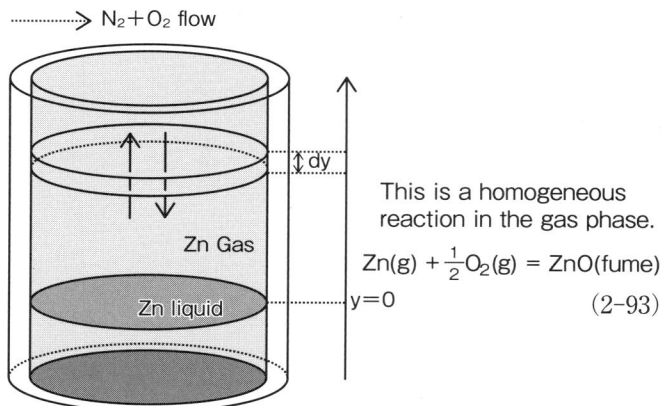

$$Zn(g) + \frac{1}{2}O_2(g) = ZnO(\text{fume}) \quad (2\text{-}93)$$

Fig. 2-7 Evaporation of Zn and reaction of Zn gas with oxygen.

Perform a mole balance on each component separately since the concentration profile of Zn can not be differential form, that of O_2 or vice versa, because of the chemical reaction.

<Equations>

① stoichimetry

$$\frac{\dot{n}_{O_2}}{A} = -\frac{1}{2}\frac{\dot{n}_{Zn}}{A} \qquad (2\text{-}94)$$

(2-93)式より(2-94)式のように表される。

② mass balance of Zn and O_2

Molar balance for Zn in the gas phase by choosing a sectional control volume as shown in Fig. 2-7.

微小領域dyにおけるインプットとアウトプットの関係は亜鉛については(2-100)式、酸素については(2-104)式のようになる。

Rate of imput of Zn $\dfrac{\dot{n}_{Zn}}{A}A$ (2-95)

Rate of output of Zn $\dfrac{\dot{n}_{Zn}}{A}A + \dfrac{d}{dy}\left(\dfrac{\dot{n}_{Zn}}{A}\right)A dy$ (2-96)

Consumption of Zn $\dot{r}_{Zn} A dy$ (2-97)
                  mole/cm³·sec   cm³

Rate of generation   0   (2-98)

Rate of accumulation   0 (steady state) (2-99)

Balance of Zn is

$$\frac{\dot{n}_{Zn}}{A}A - \left\{\frac{\dot{n}_{Zn}}{A}A + \frac{d}{dy}\left(\frac{\dot{n}_{Zn}}{A}A\right)dy\right\} - \dot{r}_{Zn} \cdot A dy = 0 \quad (2\text{-}100)$$

Similarly, balance of O_2 is

Rate of input of O_2 $-\dfrac{\dot{n}_{O_2}}{A}A$ (2-101)

Rate of output of O_2 $-\dfrac{\dot{n}_{O_2}}{A}A - \dfrac{d}{dy}\left(\dfrac{\dot{n}_{O_2}}{A}\right)A dy$ (2-102)

Rate of consumption of O_2 $\dot{r}_{O_2} A dy$ (2-103)

$$\therefore \frac{d}{dy}\left(\frac{\dot{n}_{O_2}}{A}\right) A dy - \dot{r}_{O_2} A dy = 0 \qquad (2\text{-}104)$$

③ rate of chemical reaction

The rate of the homogeneous chemical reaction; assume a second order irreversible reaction.

$$\dot{r}_{O_2} = \frac{1}{2}\dot{r}_{Zn} = k_r C_T^2 X_{Zn} \cdot X_{O_2} \qquad (2\text{-}105)$$

ここで(2-93)式の反応を、2次の均一不可逆反応と仮定すると、反応速度は(2-105)式で表される。

④ molar flux equation

$X_{Zn}, X_{O_2} \ll 1, \quad X_{N_2} \gg X_{O_2}, X_{Zn}$ (2-106)

hence neglect bulk diffusion

ガス中の酸素濃度、亜鉛蒸気濃度は窒素に比べて十分に低いため、X_{Zn}, X_{O_2} ≃0 とみなすことができる。

したがってモルフラックスの式は亜鉛と酸素について、それぞれ(2-107)、(2-108)式のようになる。

$$\frac{\dot{n}_{Zn}}{A} = -D_{Zn,N_2}C_T\frac{dX_{Zn}}{dy} + \overset{0}{\underset{\parallel}{X_{Zn}}}\left(\frac{\dot{n}_{Zn}}{A} + \frac{\dot{n}_{O_2}}{A}\right)$$
(2-107)

$$\frac{\dot{n}_{O_2}}{A} = -D_{O_2,N_2}C_T\frac{dX_{O_2}}{dy} + \underset{\underset{0}{\parallel}}{X_{O_2}}\left(\frac{\dot{n}_{Zn}}{A} + \frac{\dot{n}_{O_2}}{A}\right)$$
(2-108)

$$(\because X_{N_2} \gg X_{O_2}, X_{Zn})$$

Substitute for flux $\left(\frac{\dot{n}_{O_2}}{A}, \frac{\dot{n}_{Zn}}{A}\right)$ and reaction rate $(\dot{r}_{Zn}, \dot{r}_{O_2})$ in molar balance,

以上より、亜鉛と酸素に関する物質収支式は(2-109)、(2-110)式となる。

$$-\frac{d}{dy}\left(-D_{Zn,N_2}C_T\frac{dX_{Zn}}{dy}\right) - 2k_rC_T^2X_{Zn}\cdot X_{O_2} = 0$$
(2-109)

$$\frac{d}{dy}\left(-D_{O_2,N_2}C_T\frac{dX_{O_2}}{dy}\right) - k_rC_T^2X_{Zn}\cdot X_{O_2} = 0$$
(2-110)

To obtain X_{Zn} and X_{O_2}, we have to solve the two-second order dimensional equations that have to be solved simultaneously.

この系に関して、さらに簡略化された条件での移動速度の記述を考えてみる。

Simplification; assuming a reaction plane at y=Y, because reaction rate is very fast relative to diffusion rate.

図2-8に示すように容器下部から亜鉛蒸気、上部から酸素ガスが拡散し容器の高さy=Yにおいて(2-93)式の反応が起こると仮定する。

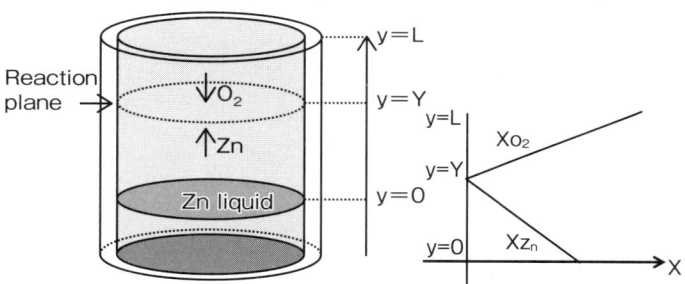

Fig. 2-8　Simplified concentration profile of Zn and oxygen.

Yより下部(0<y<Y)では酸素濃度は0、Yより上部(Y<y<L)では亜鉛蒸気濃度は0であるため、(2-109)、(2-110)式は(2-111)、(2-112)式のようになる。これを(2-113)～(2-115)式の境界条件で解くと(2-116)式で示される濃度分布が得られる。

$0<y<Y, X_{O_2}=0$

$$D_{Zn,N_2}C_T\frac{d^2X_{Zn}}{dy^2} = 0 \rightarrow \frac{dX_{Zn}}{dy} = \text{const.} \rightarrow \text{linear}$$
(2-111)

$Y<y<L, X_{Zn}=0$

$$D_{O_2,N_2}C_T\frac{d^2X_{O_2}}{dy^2} = 0 \rightarrow \frac{dX_{O_2}}{dy} = \text{const.} \rightarrow \text{linear}$$
(2-112)

B.C.
$$y=0 \quad X_{Zn}=X_{Zn,0} \quad (2\text{-}113)$$
$$y=Y \quad X_{Zn}=0, \ X_{O_2}=0 \quad (2\text{-}114)$$
$$y=L \quad X_{O_2}=X_{O_2,L} \quad (2\text{-}115)$$

solution
$$\frac{X_{Zn}}{X_{Zn,0}}=\frac{Y-y}{Y}, \quad \frac{X_{O_2}}{X_{O_2,L}}=\frac{y-Y}{L-Y} \quad (2\text{-}116)$$

$$\therefore \frac{\dot{n}_{Zn}}{A}=\frac{D_{Zn,N_2}C_T X_{Zn,0}}{Y} \quad (2\text{-}117)$$

$$\therefore \frac{\dot{n}_{O_2}}{A}=-D_{O_2,N_2}C_T \frac{X_{O_2,L}}{L-Y} \quad (2\text{-}118)$$

Y in general is unknown. Use stoichiometry to relate $\dot{n}_{O_2}$ to $\dot{n}_{Zn}$ and calculate Y.

From stoichiometry
$$\frac{\dot{n}_{O_2}}{A}=-\frac{1}{2}\frac{\dot{n}_{Zn}}{A} \quad (2\text{-}119)$$

$$\therefore \frac{D_{Zn,N_2}C_T X_{Zn,0}}{2Y}=\frac{D_{O_2,N_2}C_T X_{O_2,L}}{L-Y} \quad (2\text{-}120)$$

$$\therefore Y=\frac{L}{1+\dfrac{2D_{O_2,N_2}X_{O_2,L}}{D_{Zn,N_2}X_{Zn,0}}} \quad (2\text{-}121)$$

Substitute into flux equation
$$\therefore \boxed{\frac{\dot{n}_{Zn}}{A}=\frac{D_{Zn,N_2}C_T X_{Zn,0}}{L}\left(1+\frac{2D_{O_2,N_2}X_{O,L}}{D_{Zn,N_2}X_{Zn,0}}\right)} \quad (2\text{-}122)$$

In the absence of a chemical reaction,
$$\boxed{\frac{\dot{n}_{Zn}}{A}=\frac{D_{Zn,N_2}C_T X_{Zn,0}}{L}} \quad (2\text{-}123)$$

The flux in the presence of chemical reaction is greater than the flux in the absence of the reaction by a factor of
$$\boxed{1+\frac{2D_{O_2,N_2}X_{O,L}}{D_{Zn,N_2}X_{Zn,0}}} \quad (2\text{-}124)$$

これより、亜鉛と酸素に関する流束式はそれぞれ(2-117)、(2-118)式となる。

化学量論性の関係より、反応の位置Yは(2-121)式で与えられ、最終的に亜鉛の流束式は(2-122)式で与えられる。

ここで反応が無い場合の亜鉛の流束式は(2-123)式となるため、反応が有る場合は無い場合に比べ亜鉛の流束は(2-124)式の係数倍、大きくなると言える。

III. Unsteady State Diffusion (Fick's 2nd law) and Relating Mass Transfer

1. Fick's 2nd law

Consider a system in which concentration and flux of given species changes as a function of x,y,z and time t.

We will derive the governing differential equations by applying a mass balance to a volume element.

ある元素の濃度やモル流束が場所や時間と共に変化する系を考える。

図3-1に示す体積要素において物質収支をとり、支配方程式を導出する。

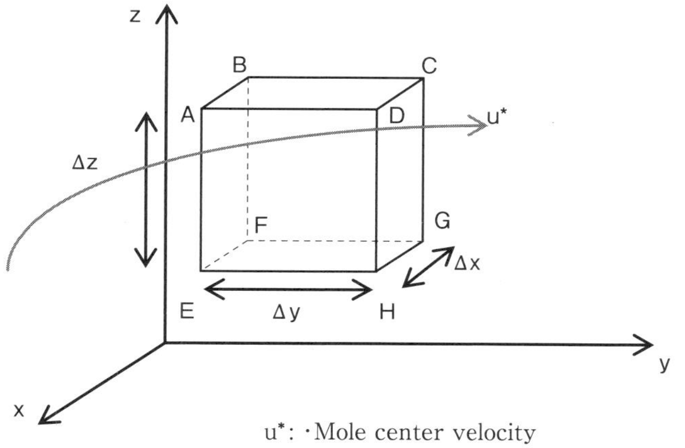

u^*: ・Mole center velocity
・Bulk diffusion and velocity due to convection.

Fig. 3-1 Element of volume to take mass balance.

ここで u^* はモル平均速度で、バルク中の拡散や対流に基づく流れである。

$$u^* = \vec{u}_x^* + \vec{u}_y^* + \vec{u}_z^* \tag{3-1}$$

$$\frac{\dot{n}_i}{A} = \frac{\dot{n}_{ix}}{A} + \frac{\dot{n}_{iy}}{A} + \frac{\dot{n}_{iz}}{A} \tag{3-2}$$

流れ、およびモル流束のx、y、z成分はそれぞれ(3-1)、(3-2)式である。

それぞれの面に入るモル速度は(3-3)～(3-5)式で表される。

Rate of input of i @BCFG $= \dfrac{\dot{n}_{ix}}{A} dydz$ (3-3)

Rate of input of i @ABFE $= \dfrac{\dot{n}_{iy}}{A} dxdz$ (3-4)

Rate of input of i @EFGH $= \dfrac{\dot{n}_{iz}}{A} dxdy$ (3-5)

それぞれの面から出るモル流量は(3-6)～(3-8)式で表される。

Rate of output of i @AEHD
$$= \dfrac{\dot{n}_{ix}}{A} dydz + \dfrac{\partial}{\partial x}\left(\dfrac{\dot{n}_{ix}}{A} dydz\right)dx \quad (3\text{-}6)$$

Rate of output of i @DCGH
$$= \dfrac{\dot{n}_{iy}}{A} dxdz + \dfrac{\partial}{\partial y}\left(\dfrac{\dot{n}_{iy}}{A} dxdz\right)dy \quad (3\text{-}7)$$

Rate of output of i @ABCD
$$= \dfrac{\dot{n}_{iz}}{A} dxdy + \dfrac{\partial}{\partial z}\left(\dfrac{\dot{n}_{iz}}{A} dxdy\right)dz \quad (3\text{-}8)$$

要素体積中で生成する速度、消費される速度、蓄積される速度はそれぞれ(3-9)、(3-10)、(3-11)式のようになる。

Rate of generation of i in volume
$$= \dot{r}_g \cdot dx \cdot dy \cdot dz \quad (3\text{-}9)$$

Rate of consumption of i in volume
$$= \dot{r}_c \cdot dx \cdot dy \cdot dz \quad (3\text{-}10)$$

Rate of accumulation of i in volume
$$= \dfrac{\partial C_i}{\partial t} \cdot dx \cdot dy \cdot dz \quad (3\text{-}11)$$

以上より濃度の時間変化は(3-12)式で表される。

ここで、$\dfrac{\dot{n}_{ix}}{A}$, $\dfrac{\dot{n}_{iy}}{A}$, $\dfrac{\dot{n}_{iz}}{A}$ には体積流れを考慮したFickの第一法則が適用される。

Therefore,
$$\dfrac{\partial C_i}{\partial t} = -\dfrac{\partial}{\partial x}\left(\dfrac{\dot{n}_{ix}}{A}\right) - \dfrac{\partial}{\partial y}\left(\dfrac{\dot{n}_{iy}}{A}\right) - \dfrac{\partial}{\partial z}\left(\dfrac{\dot{n}_{iz}}{A}\right) + \dot{r}_g - \dot{r}_c \quad (3\text{-}12)$$

where $\dfrac{\dot{n}_{ix}}{A} = -D_{i,m} C_i \dfrac{\partial X_i}{\partial x} + u_x^* X_i C_T$ (3-13)

$\dfrac{\dot{n}_{iy}}{A} = -D_{i,m} C_i \dfrac{\partial X_i}{\partial y} + u_y^* X_i C_T$ (3-14)

$\dfrac{\dot{n}_{iz}}{A} = -D_{i,m} C_i \dfrac{\partial X_i}{\partial z} + u_z^* X_i C_T$ (3-15)

X_i : molar fraction of i ($= \dfrac{C_i}{C_T}$)

u_x^*, u_y^*, u_z^* are the three components of fluid velocity

Substituting for $\dot{n}_{ix}$, $\dot{n}_{iy}$, $\dot{n}_{iz}$, C_i

最終的に、(3-16)式のように集約される。

$$-\frac{\partial}{\partial x}\left(-D_{i,m}C_T\frac{\partial X_i}{\partial x}\right)-\frac{\partial}{\partial y}\left(-D_{i,m}C_T\frac{\partial X_i}{\partial y}\right)$$

$$-\frac{\partial}{\partial z}\left(-D_{i,m}C_T\frac{\partial X_i}{\partial z}\right) \quad \text{①}$$

diffusion term

$$-\frac{\partial}{\partial x}(u_x^* C_T X_i)-\frac{\partial}{\partial y}(u_y^* C_T X_i)-\frac{\partial}{\partial z}(u_z^* C_T X_i) \quad \text{②}$$

convection term

$$+\dot{r}_g - \dot{r}_c \quad \text{③}$$

$$=\frac{\partial}{\partial t}(X_i C_T) \tag{3-16}$$

<Simplification>

① $D_{i,m}$ is not a function of x, y, z. C_T is constant,

$$C_T D_{i,m}\left(\frac{\partial^2 X_i}{\partial x^2}+\frac{\partial^2 X_i}{\partial y^2}+\frac{\partial^2 X_i}{\partial z^2}\right) \tag{3-17}$$

② Taking account of eq. of continuity

$$\frac{\partial u_x}{\partial x}+\frac{\partial u_y}{\partial y}+\frac{\partial u_z}{\partial z}=0 \tag{3-18}$$

therefore

$$(u_x^*\frac{\partial}{\partial x}(C_T X_i)+C_T X_i\frac{\partial u_x}{\partial x})$$

$$+(u_y^*\frac{\partial}{\partial y}(C_T X_i)+C_T X_i\frac{\partial u_y}{\partial y})$$

$$+(u_z^*\frac{\partial}{\partial z}(C_T X_i)+C_T X_i\frac{\partial u_z}{\partial z})$$

$$=u_x^*\frac{\partial C_T X_i}{\partial x}+u_y^*\frac{\partial C_T X_i}{\partial y}+u_z^*\frac{\partial C_T X_i}{\partial z}$$

$$+C_T X_i\cancel{\left(\frac{\partial u_x}{\partial x}+\frac{\partial u_y}{\partial y}+\frac{\partial u_z}{\partial z}\right)} \tag{3-19}$$

<Note> Similarity to energy transport

$$\frac{\partial}{\partial x}(k\frac{\partial T}{\partial x})+\frac{\partial}{\partial y}(k\frac{\partial T}{\partial y})+\frac{\partial}{\partial z}(k\frac{\partial T}{\partial z})+u\frac{\partial T}{\partial x}+v\frac{\partial T}{\partial y}+w\frac{\partial T}{\partial z}$$

$$+\dot{q}_g-\dot{q}_c=\rho C_p\frac{\partial T}{\partial t} \tag{3-20}$$

ここで簡略化を行う。

① $D_{i,m}$ および C_T は、定数とみなす。

② 連続の式（(3-18)式）を考慮すると(3-19)式のようになる。

③ 化学反応は生じない。

③ No chemical reactions, $\dot{r}_g, \dot{r}_c = 0$ (3-21)

$$\frac{\partial X_i}{\partial t} = D_{i,m}\left(\frac{\partial^2 X_i}{\partial x^2} + \frac{\partial^2 X_i}{\partial y^2} + \frac{\partial^2 X_i}{\partial z^2}\right) - \left(u_x^*\frac{\partial X_i}{\partial x} + u_y^*\frac{\partial X_i}{\partial y} + u_z^*\frac{\partial X_i}{\partial z}\right) \quad (3\text{-}22)$$

④ バルク流れは無く、x方向のみの拡散を考える。

　以上のようにして、(3-23)式で示されるFickの第二法則（x方向）の基礎式を得る。

④ No bulk flow & unidirectional diffusion

$$\boxed{D_{i,m}\frac{\partial^2 X_i}{\partial x^2} = \frac{\partial X_i}{\partial t}} \quad (3\text{-}23) \sim \text{ Fick's 2}^{nd} \text{ law}$$

To obtain distribution of Xi in a metallurgical reaction the unsteady state diffusion eq. has to be solved together with the continuity eq. and Navier-Stokes eqs. to determin u^*, Xi, $\dot{n}_i$.

2. Mass transfer coefficient and diffusion coefficient

$$\boxed{\frac{\dot{n}_i}{A} = k_i(C_i - C_i^*)} = (\text{conductance}) \times (\text{driving force})$$

$$\left[\begin{array}{l} \dfrac{\dot{n}_i}{A} \ (\text{mole/m}^2 \cdot \text{sec}) \\ C_i, \ C_i^* \ (\text{mole/m}^3) \end{array}\right]$$

k_i mass transfer coefficient (m/sec)

(3-24)

(3-24)式に示すように、モル流束はコンダクタンス（抵抗の逆数）と駆動力（濃度差）の積で表される。このコンダクタンスを物質移動係数と呼ぶ。

<Note>

Role similarity

(1) Ohm's law
  $i = \Delta V/R$ (3-25)

(2) Heat transfer
  $\dfrac{q}{A} = k\Delta T$ (3-26)

(3) Mass transfer
  $\dfrac{\dot{n}_i}{A} = k_i(\Delta C)$ (3-27)

flux = **conductance** × driving force
(current) (voltage difference)
(heat)   (temp.   〃)
(mass)   (concentration 〃)

Instead of solving the unsteady state diffusion equation, a mass transfer coefficient is often used to calculate fluxes.

Factors affecting the mass transfer coefficient :
(1) fluid properties
(2) system geometry
(3) level of turbulence
(4) diffusion coefficient

モル流速を求めるに当たっては、非定常拡散の式を解くのではなく、物質移動係数がよく用いられる。物質移動係数に影響するプロセス要素としては、流体の物性、系の形状、流れの状態（乱流の状態）、拡散速度（拡散係数）などがある。

Methods of determining the mass transfer coefficient ;
(1) measurement on process models, pilot plant or prototype.
(2) dimensionless correlations from literature.
  Sherwood number = f (Reynolds number, Schmidt

物質移動係数を求める手段としては、
(1) パイロットプラント等での実験測定値から
(2) 文献で報告されている無次元数の関係式から

where Sherwood number : $\dfrac{\text{total mass transfer}}{\text{molecular diffusivity}}$

Schmidt number : $\dfrac{\text{momentum diffusivity}}{\text{molecular diffusivity}}$

Reynolds number : $\dfrac{\text{innertia forces}}{\text{viscous forces}}$

(3) 熱・物質移動解析からの推定

(3) analysis between heat & mass transfer.

e.g. fluid flowing over a flat plate

heat transfer $N_u = f(Re, Pr)$ (3-28)

N_u: $\dfrac{\text{total heat transfer}}{\text{thermal diffusivity}}$ Pr: $\dfrac{\text{momentum diffusivity}}{\text{thermal diffusivity}}$

mass transfer $Sh = g(Re, Sc)$ (3-29)

(4) 物質移動モデル（境膜モデル、浸透モデル）などがある。

(4) calculation of mass transfer coefficient from models.

Mass transfer models

a) film theory

b) penetration theory

2.1 Film theory

境膜理論とFickの第二法則の適用について説明する。

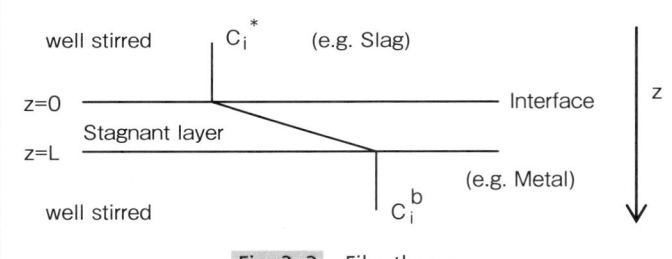

Fig. 3-2 Film theory.

In this model stirred fluid such as a metal is in contact with a slag.

In the vicinity next to the interface is a stagnant layar and mass transfer of the species takes place solely by molecular diffusion. This represents the total resistance to mass transfer is in the fluid, since it is the slowest transport step.

<Assumptions>

1) No generation and no consumption.
   $\dot{r}_g = 0, \dot{r}_c = 0$ in diffusion equation.

2) Steady state in the stagnant layer.
   $\frac{\partial C_i}{\partial t} = 0$

3) Since layer is stagnant X_i is small and $u^* = 0$.

4) Unidirectional in z direction only.

<Equation>

$$D_{i,m} \frac{\partial^2 C}{\partial z^2} = 0 \quad (3\text{-}30)$$

$$\frac{\partial C}{\partial z} = \alpha, \text{ const.} \quad (3\text{-}31)$$

<Solution>

$$C = \frac{C_i^b - C_i^*}{L} z + C_i^* \quad (3\text{-}32)$$

$$\boxed{\frac{\dot{n}_i}{A} = -D_{i,m} \frac{\partial C}{\partial z}} \quad (\because \text{medium stagnant})$$

$$= -D_{i,m} \frac{C_i^b - C_i^*}{L} \quad (3\text{-}33)$$

$$\boxed{k_i = \frac{D_{i,m}}{L}} \longleftarrow \text{film theory} \quad (3\text{-}34)$$

1) 生成や消費は無い
2) 定常状態
3) 淀んでいるため流速は0、成分濃度は希薄である
4) 一方向の物質移動

定常状態であるため、(3-23)式は(3-30)式のようになる。
図3-2に示すように $z=0$ で $C_i = C_i^*$、$z=L$ で $C_i = C_i^b$ とする。

(3-32)式のように直線関係の濃度勾配の下、モル流束は(3-33)式のように表される。
なお、右辺のマイナスは、移動方向を示すものであり、これを考慮して、(3-33)式を(3-24)式と対比させると物質移動係数 k_i は $(D_{i,m}/L)$ に相当することがわかる。Lは境膜厚みである。

<Problems>

1) The concept of a stagnant layer is not very realistic for most metallurgical systems.

2) It is not easy to estimate L except for very simple cases, in which the velocity distribution is unknown.

2.2 Penetration theory

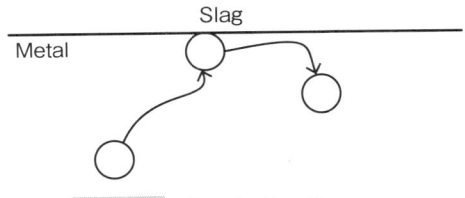

Fig. 3-3 Penetration theory

続いて、浸透理論としてのFickの第二法則の適用について述べる。

この理論は乱流が十分に発達した液相をベースとしている。図3-3に示すように液体の要素（渦）はバルク→界面→バルクへと移動する。この要素（渦）が界面にあってバルクに戻るまでの間、溶質の非定常の移動が進行すると考える。

The theory considers a fully turbulent fluid. The turbulence is such that packets of fluid or eddies travel from bulk to the interface.

Unsteady state mass transfer of solute takes place from the eddy during the time the eddy is at the interface and then it is swept back into the fluid and a new eddy takes place.

(3-35)式(=(3-23)式)そのものを(3-36)～(3-38)式の境界条件下で、ラプラス変換で解くと(3-39)式が得られる。erfcは相補誤差関数であり(3-40)式で定義される。

$$\boxed{\frac{\partial C_i}{\partial t} = D_{i,m} \frac{\partial^2 C_i}{\partial z^2}} \quad (3\text{-}35)$$

B.C.
i) $z=0, t>0, C_i = C_i^*$ (3-36)

ii) $z \geq 0, t=0, C_i = C_i^b$ (3-37)

iii) $z \to \infty, t \geq 0, C_i = C_i^b$ (3-38)

Laplace transformation can be used to solve the equation.

$$\frac{C_i - C_i^b}{C_i^* - C_i^b} = \text{erfc}\left\{\frac{z}{2\sqrt{D_{i,m} t}}\right\} \quad (3\text{-}39)$$

$$\text{erfc}(z) = 1 - \frac{2}{\sqrt{\pi}} \int_0^{f(z)} e^{-\xi^2} d\xi \quad (3\text{-}40)$$

we are interested in

$$\frac{\dot{n}_i}{A} = -D_{i,m} \left.\frac{\partial C_i}{\partial z}\right|_{z=0} \quad (3\text{-}41)$$

(assuime Xi is small, so no bulk diffusion)

$$\frac{\partial C_i}{\partial z} = \frac{\partial}{\partial z}\left\{(C_i^* - C_i^b)\text{erfc}\left(\frac{z}{2\sqrt{D_{i,m}t}}\right) + C_i^b\right\} \quad (3\text{-}42)$$

$$= \frac{(C_i^* - C_i^b)}{\sqrt{D_{i,m}\pi t}} \cdot e^{-\frac{z^2}{4D_{i,m}t}} \quad (3\text{-}43)$$

(3-39)式は(3-43)式のように変形される。

z=0におけるモル流束は(3-41)式で与えられるため、(3-43)式はさらに(3-44)式のように変形される。

(3-44)式を(3-24)式と対比すると物質移動係数は(3-45)式となることがわかる。

$$\therefore \boxed{\frac{\dot{n}_i}{A} = \sqrt{\frac{D_{i,m}}{\pi t}} (C_i^* - C_i^b)} \quad (3\text{-}44)$$

This is the instantaneons flux of i at the z=0 plane.

Compare with

$$\frac{\dot{n}}{A} = k_i (C_i^* - C_i^0) \quad (3\text{-}24)$$

$$\boxed{k_i = \sqrt{\frac{D_{i,m}}{\pi t}}} \quad (3\text{-}45)$$

Average mass transfer rate of it when an eddy remaines for a time (t_e).

$$\boxed{\overline{\frac{\dot{n}_i}{A}} = \frac{1}{t_e}\int_0^{t_e}\left(\frac{\dot{n}_i}{A}\right)dt = \frac{1}{t_e}\int_0^{t_e}(C_i^* - C_i^b)\frac{\sqrt{D_{i,m}}}{\sqrt{\pi}}t^{-\frac{1}{2}}dt}$$
$$\boxed{= (\overline{C_i^*} - \overline{C_i^b}) \cdot 2\sqrt{\frac{D_{i,m}}{\pi \cdot t_e}}} \quad (3\text{-}46)$$

$$\boxed{\overline{k}_i = 2\sqrt{\frac{D_{i,m}}{\pi \cdot t_e}}} \quad \longleftarrow \text{penetration theory} \quad (3\text{-}47)$$

ここで、流れが時間t_eの間、反応に関与すると考えると、その時間平均での物質移動係数は(3-47)式のようになる。
すなわち、物質移動係数は最終的に(3-46)式のようになるが、この考え方は浸透理論と呼ばれている。

How to determine (t_e) and ($\overline{k}_i$) for specific cases.

この物質移動係数$\overline{k}_i$はその性質上、流れ場、すなわち滞留時間t_eの影響を受けるため、反応場や流れ場によって種々の仮定に基づき決定される。

1) Babbles rising through liquid

1) 液体中を気泡が上昇する場合
$$t_e = \frac{h}{u}$$
ここで h は spherical cap の高さである。

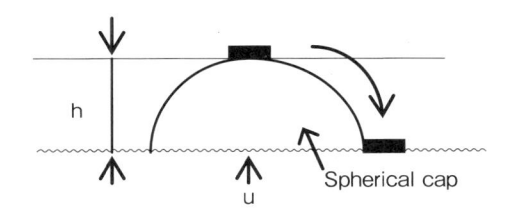

Fig. 3-4 Babbles rising through liquid.

velocity of rise of spherical cap babbles

$$u = 1.02\sqrt{gr_e} \quad (\text{cm/s}) \quad (3\text{-}48)$$

r_e : radius of sphere of some volume as spherical cap

t_e : time the element of fluid is in contact with bubble as it moves from the frontal stagnation point to the trailing edge of the bubble

$$\boxed{t_e = \frac{h}{u}} \quad (3\text{-}49)$$

2) 液体が気泡により攪拌される場合
$$t_e = \frac{1}{f}$$
ここでfは気泡の吹き込み頻度である。

2) Bubbles stirred interface

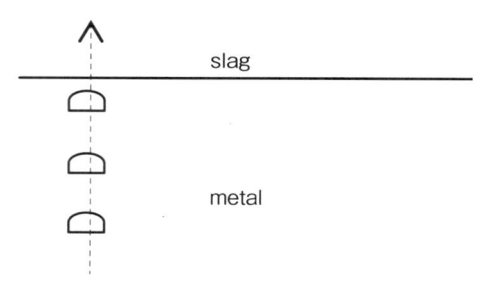

Fig. 3-5 Bubbles stirring interface.

frequency of bubbles rising through the interface : f

$$\bar{k}_i = 2\sqrt{\frac{D_i f}{\pi}} \qquad (3\text{-}50)$$

experimentally it has been shown that

$$\bar{k}_i \propto f^{0.5} \cdot D_{i,m}^{0.5} \cdot V^{0.42} \qquad (3\text{-}51)$$
$\qquad\qquad\qquad\qquad$ ⌣bubble volume

$$t_e = \frac{1}{f} \qquad (3\text{-}52)$$

3) 固体の粒子が液体中を落下する場合
$$t_e = \frac{d}{u_d}$$
ここでu_dは粒子の上昇（下降）速度であり、たとえばストークスの式などから求められる。

3) Drops falling through liquid

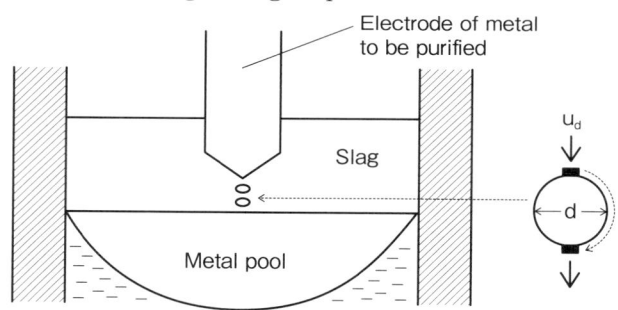

Fig. 3-6 Metal drops falling through liquid slag in electro slag remelting process.

1) for small drop Re < 1 (0.5mm)

the drops behave as rigid spheres with no internal circulation

$$\bar{k}_i = 2\sqrt{\frac{D_{i,m} u_d}{\pi d}} \qquad (3\text{-}53)$$

$$t_e = \frac{d}{u_d} \qquad (3\text{-}54)$$

2) for large drop with internal circulation
$$Sh = 50 + 0.085\, Re\, Sc^{0.7} \qquad (3\text{-}55)$$

4) Induction stirred melt

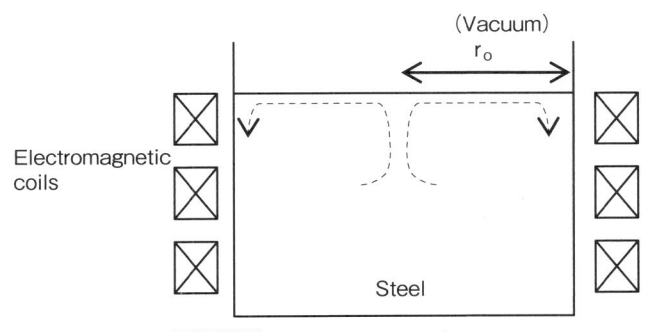

Fig. 3-7 Induction stirred melt.

In this case it is supposed that an eddy penetrats to the surface at the center of the bath and travels acrocss the bath surface with velocity u. Unsteady state diffusion takes place within the eddy. The average mass transfer coefficient is

$$\overline{k}_i = \frac{4}{\sqrt{3\pi}} \left(\frac{uD_{i,m}}{r_0} \right)^{\frac{1}{2}} \tag{3-56}$$

In the case $u_{r0} \cdot r_0 =$ const. (u_{r0} : velocity at outer radius)

$$\overline{k}_i = \left(\frac{8}{\pi} \right)^{\frac{1}{2}} [(u_{ro} \cdot r_o)D_{i,m}]^{\frac{1}{2}} \tag{3-57}$$

5) Top blown jet

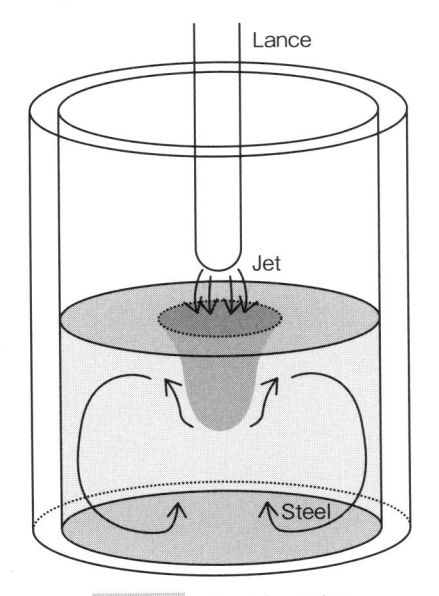

Fig. 3-8 Top blown jet.

4) 電磁攪拌されている溶解炉でのスラグ／メタル反応

$$\overline{k}_i = \frac{4}{\sqrt{3\pi}} \left(\frac{uD_{i,m}}{r_0} \right)^{\frac{1}{2}}$$

r_0は炉の半径

炉の外周における流速がu_{ro}で積$u_{ro} \cdot r_o$が一定の場合、

$$\overline{k}_i = \left(\frac{8}{\pi} \right)^{\frac{1}{2}} [(u_{ro} \cdot r_o)D_{i,m}]^{\frac{1}{2}}$$

5) 上吹きジェット攪拌によるスラグ／メタル反応

$$\overline{k}_i = \frac{4}{\sqrt{3\pi}} \left(\frac{uD_{i,m}}{r_0} \right)^{\frac{1}{2}}$$

(3-56)式と同等の式で与えられる。

$$\bar{k}_i = \frac{4}{\sqrt{3\pi}} \left(\frac{uD_{i,m}}{r_o} \right)^{\frac{1}{2}} \qquad (3\text{-}58)$$

This is considered to be the same as the induction stirred melt.

Alternative mass transfer models,
- film theory $k_i \propto D_{i,m}^{1.0}$
- penetration theory $k_i \propto D_{i,m}^{0.5}$

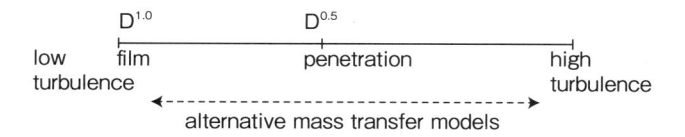

```
         D^1.0           D^0.5
         |                 |                    |
low      film           penetration            high
turbulence                                     turbulence
         <------------------------------------>
              alternative mass transfer models
```

$k_i \propto D_{i,m}^n \qquad 0 < n < 1.0$

n : depends on the degree of turbulence

IV. General Formulation of Mass Transfer in Batch Process

1. Basic equations for batch process

The general aim in any process control is to find out how the concentration of given species varies as a function of time and position in a metallurgical reactor.

Consider a batch process as an example (see Fig. 4-1). To formulate the mathematical model of the problem we have the following equations.

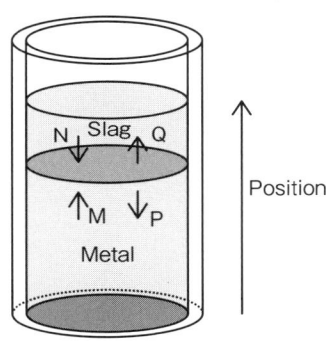

Fig. 4-1 Batch reactor.

1) Molar balance equations for species
2) Molar flux equations
3) Stoichiometry equations
4) Local equilibrium (assuming the reaction at the interface is instantaneous)

The reaction being interested is,

$$mM + nN = pP + qQ \qquad (4\text{-}1)$$

We want to find out how the concentration of M changes with time in the bulk, $C_M^b(t)$.

(1) Molar balance for M in the bath

Assume bath is well mixed and gradient of M is confined

プロセス制御を行う上で最も重要な項目の一つは、反応容器内にある溶鋼の所定の位置の、所定の成分濃度の時間変化を予測することである。

図4-1に示すバッチプロセスを例に、予測のための定式化について紹介する。これまでに学んだ物質移動の基礎をベースに、成分濃度の時間変化を推定するための数式モデルを作成するために、次の4項目についての関係式を得る必要がある。

1) 種々の成分についてのモルバランス
2) モル流束
3) 化学量論性を考慮したモル流束式
4) 反応平衡の状態にある界面における種々の溶質成分の濃度バランス

たとえば(4-1)式の反応が界面で進行している系において、溶鋼バルクの成分

Mの時間変化を知ろうとすると、Mに関するモルバランスの式（(4-2) 式）が必要である。

同様の関係式をN、P、Qについて得る必要がある。

to a small reagion near the slag/metal interface.

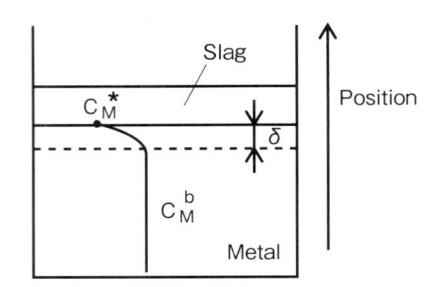

Fig. 4-2　Concentration profile of M in metal.

Applying the balance,

Rate of input of M to metal bath $=0$

Rate of output of M from metal bath $=\dfrac{\dot{n}_M}{A}\cdot A$

Rate of generation & consumption of M $=0$

Rate of accumulation of M $=V_b\dfrac{dC_M^b}{dt}$

$$\therefore\ -\dfrac{\dot{n}_M}{A}A=V_b\dfrac{dC_M^b}{dt} \tag{4-2}$$

where　$\dot{n}_M$　; mass flux of M

　　　　A　; interface area

　　　　V_b　; volume of metal bath

　　　　C_M^b　; concentration of M in metal bath

　　　　t　; time

Same type of equations for N, P, and Q.

M、Nについてのモル流束の式は (4-3) 式，(4-4) 式のようになる。同様の関係式をP、Qについて得る必要がある。反応界面における反応平衡より、M、N、P、Qの濃度は (4-5) 式によって関係づけられる。

(2) Molar flux equations

$$\dfrac{\dot{n}_M}{A}=k_M(C_M^b-C_M^*) \tag{4-3}$$

$$\dfrac{\dot{n}_N}{A}=-k_N(C_N^b-C_N^*) \tag{4-4}$$

where　k_M　; mass transfer coefficient for M

　　　　C_M^*　; concentration of M at the interface.

Same type of equations for P, Q.

(3) Chemical equation at the interface

Because metallurgical reactions are instantaneous,

$$K = \frac{(C_P^*)^p (C_Q^*)^q}{(C_M^*)^m (C_N^*)^n} \tag{4-5}$$

where K ; effective equilibrium constant

(4) Stoichiometry equations

$$n\frac{\dot{n}_M}{A} = -m\frac{\dot{n}_N}{A} \tag{4-6}$$

$$p\frac{\dot{n}_M}{A} = -m\frac{\dot{n}_P}{A} \tag{4-7}$$

$$q\frac{\dot{n}_M}{A} = m\frac{\dot{n}_Q}{A} \tag{4-8}$$

　M、N、P、Qそれぞれのモル流束は化学量論性を考慮して（4-6）～（4-8）式のように関係づけられる。

2. Linear thermodynamics

応用熱力学と物質移動論が組み合わさった問題の解法について説明する。

最初に、ここで言う"Linear thermodynamics"とは本来の「線形熱力学」とは異なり、一般的な解を持つことが出来る物質移動と反応平衡が組み合わさった事例を指していることを注記しておく。

図4-3に示すような2液相（たとえば上部はスラグ、下部はメタル）の系を考える。

ここで、2相間の界面積をA、上部液層の体積をV_s、下部液層の体積をV_bとする。

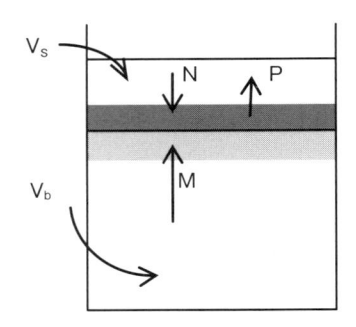

Fig. 4-3 Mass transfer and reaction at the interface of double liqued system (1).

$$M + nN \rightarrow P \qquad (4\text{-}9)$$

(4-9)式に示すように、下部液層中の成分Mの1モルと上部液層中の成分Nのnモルが、界面で反応し上部液層中に成分Pの1モルが生成する反応を考える。

M、N、Pそれぞれのバルクでの濃度をC_M^b、C_N^b、C_P^bとし、界面での濃度をC_M^*、C_N^*、C_P^*で表す。

この反応は瞬時に進行し、成分Nの界面からの移動は律速段階で無いという仮定が設けられている。

すなわち、成分Nについては(4-10)式の関係が成り立つ。

<Assumptions>

a) The reaction is instantaneous.
b) Transport of N is not rate controlling.

$$\Downarrow$$

$$C_N^* \simeq C_N^b \qquad (4\text{-}10)$$

<Equations>

ここで、ある時間 t におけるC_M^bの濃度を予測してみよう。

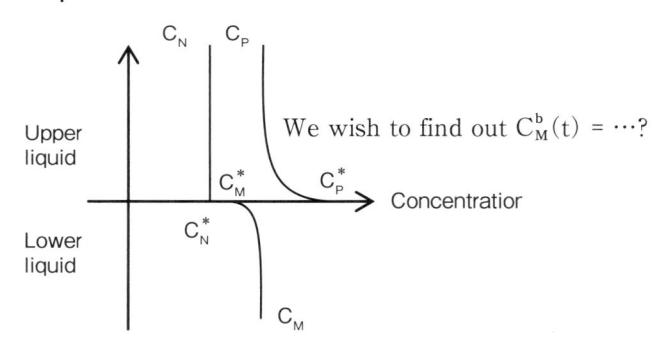

We wish to find out $C_M^b(t) = \cdots$?

Fig. 4-4 Concentration profile in upper liquid pool and lower liquid pool.

①時間t後のMに関する収支は(4-11)式のようになる。

① molar balance

$$-\frac{\dot{n}_M}{A}A = V_b \frac{\partial C_M^b}{\partial t} \qquad (4\text{-}11)$$

② molar flux

$$\frac{\dot{n}_M}{A} = k_M(C_M^b - C_M^*) \qquad (4\text{-}12)$$

$$\frac{\dot{n}_P}{A} = k_P(C_P^* - C_P^b) \qquad (4\text{-}13)$$

③ local equilibrium at interface

$$K = \frac{C_P^*}{C_M^* \cdot (C_N^b)^n} \qquad (4\text{-}14)$$

④ stoichiometry

$$\frac{\dot{n}_M}{A} = \frac{\dot{n}_P}{A} \qquad (4\text{-}15)$$

from eq. (4-14)

$$C_P^* = K \cdot C_M^* \cdot (C_N^b)^n \qquad (4\text{-}16)$$

substitute eq.(4-16) in eq.(4-13)

$$\frac{\dot{n}_P}{A} = k_P(K \cdot C_M^* \cdot (C_N^b)^n - C_P^b) \qquad (4\text{-}17)$$

$$\therefore \frac{\dot{n}_M}{A} = k_P(K \cdot C_M^* \cdot (C_N^b)^n - C_P^b) \qquad (4\text{-}18)$$

from eq.(4-12) and eq.(4-18)

$$k_M(C_M^b - C_M^*) = k_P(K \cdot C_M^* \cdot (C_N^b)^n - C_P^b) \qquad (4\text{-}19)$$

$$\therefore C_M^* = \frac{k_M C_M^b + k_P \cdot C_P^b}{k_M + k_P \cdot K(C_N^b)^n} \qquad (4\text{-}20)$$

$$\boxed{\therefore \frac{\dot{n}_M}{A} = k_M\left(C_M^b - \frac{k_M C_M^b + k_P \cdot C_P^b}{k_M + k_P \cdot K(C_N^b)^n}\right) \qquad (4\text{-}21)}$$

$$= \frac{1}{\dfrac{1}{k_M} + \dfrac{1}{k_P \cdot K \cdot (C_N^b)^n}}\left(C_M^b - \frac{C_P^b}{K(C_N^b)^n}\right) \qquad (4\text{-}22)$$

$$\boxed{k_{ov} = \frac{1}{\dfrac{1}{k_M} + \dfrac{1}{k_P \cdot \fbox{K} \cdot (C_N^b)^n}} \qquad (4\text{-}23)}$$

↳ The thermodynamics influences the kinetics of the process

② MおよびPに関するモル流束の式は、それぞれ(4-12)、(4-13)式となる。

③ 界面における反応は速く、(4-14)式の平衡が成り立っている。

④ MとPの間の化学量論式は(4-15)式となる。

まず(4-14)式と(4-13)式から(4-17)、(4-18)式を得る。

(4-12)式と(4-18)式より(4-19)式を得る。

これよりC_M^*は(4-20)式となり、Mのモル流束を整理すると(4-22)式のようになる。

( )内が駆動力、その係数がコンダクタンス、すなわちこの系の総括物質移動係数k_{ov}となる。

(4-23)式に示すように総括物質移動係数に平衡定数Kが含まれる事は、このプロセスにおいて熱力学が物質移動に関与していることを理論的に物語るものである。

2. Linear thermodynamics

ここでMとPの間の物質収支を取ると(4-24)式のようになり、バルクでの成分Pの濃度は(4-25)式で表される。

ここでV_bはメタルの体積、V_sはスラグの体積、また下部添え字の0はt=0における濃度を表す。

これを(4-22)式に代入すると、(4-26)式が得られる。

さらにC_M^bで整理して(4-27)式を得る。

(4-11)式から$\dfrac{dC_M^b}{dt}$は(4-28)式のように表され、t=0($C_{M,0}^b$)からt_f($C_{M,f}^b$)までを積分して、最終的に(4-30)式を得る。

時間t_fの時のMの濃度は$C_{M,f}^b$と求まる。

平衡定数Kが物質移動に関与していることを先に述べたが、総括物質移動係数を示す(4-23)式から、Kの大小が律速課程に影響する点も忘れてはいけない。

If we apply a molar balance for M and P,

$$V_b(C_{M,0}^b - C_M^b) = V_s(C_P^b - C_{P,0}^b) \tag{4-24}$$

$$\therefore C_P^b = \frac{V_b}{V_s}(C_{M,0}^b - C_M^b) + C_{P,0}^b \tag{4-25}$$

substitute eq.(4-22)

$$\frac{\dot{n}_M}{A} = k_{ov}\left[C_M^b - \frac{V_b}{V_s K(C_N^b)^n}(C_{M,0}^b - C_M^b) - \frac{C_{P,0}^b}{K(C_N^b)^n} \right] \tag{4-26}$$

$$= k_{ov}\left[\underbrace{\left(1 + \frac{V_b}{V_s K(C_P^b)^n}\right)}_{\alpha} C_M^b - \underbrace{\left(\frac{V_b}{V_s K(C_N^b)^n} C_{M,0}^b + \frac{C_{P,0}^b}{K(C_N^b)^n}\right)}_{\beta} \right] \tag{4-27}$$

from eq.(4-11)

$$\frac{dC_M^b}{dt} = -\frac{A}{V_b} k_{ov}[\alpha C_M^b + \beta] \tag{4-28}$$

$$\frac{A k_{ov}}{V_b} \int_0^{t_f} dt = -\int_{C_{M,0}^b}^{C_{M,f}^b} \frac{dC_M^b}{\alpha C_M^b + \beta} \tag{4-29}$$

$$\boxed{\frac{A}{V_p} k_{ov} t_f = \frac{1}{\alpha} \ln\left(\frac{\alpha C_{M,0}^b + \beta}{\alpha C_{M,f}^b + \beta}\right)} \tag{4-30}$$

The effect of the magnitude of K on the rate limiting will be discussed as follows;

$$\boxed{k_{ov} = \frac{1}{\dfrac{1}{k_M} + \dfrac{1}{k_p(C_N^b)^n K}}} \tag{4-23}$$

<Important obserbation>

(1) K affects k

The thermodynamics influences the kinetics of the process since K appears in the expression of k_{ov} and in the overall driving force.

(2) If we examine the overall mass transfer coefficient k_{ov}, we can comment on which step is the rate controlling in the process.

1) when K is large, which suggests a transport of M is rate controlling.

$$\frac{1}{k_M} \gg \frac{1}{k_P(C_N^b)^n K} \tag{4-31}$$

2) when K is small, which suggests a transport of P is rate controlling.

$$\frac{1}{k_P(C_N^b)^n K} \gg \frac{1}{K_M} \tag{4-32}$$

1) Kの値が非常に大きい場合、(4-31)式のようになり、$k_{ov}=k_M$ すなわちMの移動が律速段階となる。

2) 一方、Kの値が非常に小さい場合、(4-32)式のようになり、Pの移動が律速段階となる。

2. Linear thermodynamics

"Nonlinear thermodynamics"という表現は前節とは逆に、一般的な解を持つことが出来ない物質移動と反応平衡が組み合わさった事例を指すものである。

ここでは溶鋼の脱炭反応におけるCOガス生成反応を取り上げる。

(4-33)式で示されるこの反応に関しては、溶鋼中炭素および酸素のCO気泡表面への移動が反応抵抗となりうると考えられるが、精錬のほとんどの期間においては溶鋼中炭素濃度が酸素濃度に比べて高いため、溶鋼中炭素の移動抵抗は無視できると言われている。このことを右記の[a]～[d]の条件下で理論的に検証してみよう。

3. Nonlinear thermodynamics
(Rate controlling steps in steelmaking)

In the experiment of carbon removal (as CO gas) from steel, two of the resistances to decarburization are the transfer of dissolved carbon and oxygen atoms through the metal bath to the surface of CO bubbles which have nucleated heterogeneously on the vessel lining. At the bubble-metal interface the carbon and oxygen react instantaneously to form CO.

$$C_{(metal)} + O_{(metal)} = CO_{(gas)} \qquad (4-33)$$

It has been frequently suggested that the carbon transfer resistance is negligible compared to that of oxygen for most of the refining period since normally the bulk concentration of carbon is greater than that of oxygen. Show theoretically the conditions which must apply for this to be so. It may be helpful to recall that the carbon transfer will be negligible so long as the carbon concentrations in the bulk metal and at the metal-bubble interface are nearly equal.

Using the data below, would you expect decarburization of a metal bath at 1600°C containing 1 wt % carbon and 0.1 wt % oxygen to be controlled solely by oxygen transfer or by both carbon and oxygen transfer?

[a] Equilibrium constant $K = \dfrac{P_{co}}{[C][O]}$
$= 500$ atm wt %$^{-2}$ at 1600 °C

[b] Pressure inside the CO bubble (assumed constant)
$= 1.5$ atm.

[c] Mass transfer coefficient of oxygen in the bath (k_o)
$= 4(10^{-2})$ cm sec^{-1}

[d] Mass transfer coefficient of carbon in the bath (k_c)
$= 5.3(10^{-2})$ cm sec^{-1}

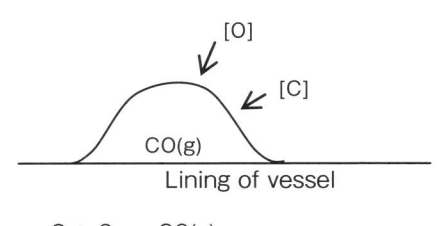

```
      [O]
       ↓ ↙ [C]
     ⌒⌒⌒
    ( CO(g) )
   Lining of vessel
```

C + O → CO(g)

Fig. 4-5 Generation of CO gas bubble on the lining of vessel.

<Assumptions>

a) Rapid reaction.
$$K = \frac{P_{co}^*}{C_c^* C_o^*} \sim C_c^* = \frac{P_{co}^b}{C_o^* K} \quad (4\text{-}34)$$

b) Carbon transfer is not a rate controlling step.
$$C_c^* \simeq C_c^b \quad (4\text{-}35)$$

<Equations>

① molar balance

→ unnecessary because not interested in $C_c^b(t)$

② molar flux
$$\frac{\dot{n}_c}{A} = k_c(C_c^b - C_c^*) \quad (4\text{-}36)$$

$$\frac{\dot{n}_o}{A} = k_o(C_o^b - C_o^*) \quad (4\text{-}37)$$

③ local equilibrium at the interface
$$K = \frac{P_{co}^*}{C_c^* C_o^*} = \frac{P_{co}^b}{C_c^* C_o^*} \quad (4\text{-}38)$$

(CO transfer in a pure CO bubble cannot be rate limiting)

④ stoichiometry
$$\frac{\dot{n}_c}{A} = \frac{\dot{n}_o}{A} \quad (4\text{-}39)$$

Procedure : eliminate C_o^* to obtain a relationship between $C_c^*, C_c^b, k_c, k_o, K$ etc. then derive condition that makes $C_c^* \simeq C_c^b$.

from eq.(4-36), eq.(4-37) and eq.(4-39)
$$k_c(C_c^b - C_c^*) = k_o(C_o^b - C_o^*) \quad (4\text{-}40)$$

図4-5はメタル／ライニング界面におけるCOガス発生とそのサイトに向かう溶鉄中の炭素と酸素を示している。

界面で生じる反応はきわめて速く、平衡状態((4-34)式)にあるとみなすことが出来る。
また、炭素の移動は律速課程にならない場合(4-35)式が成り立つ。

考察のアプローチは以下の通りである。
① マスバランスの式は、ここでは使用しない。
② モル流束の式はCとOについてそれぞれ(4-36)、(4-37)式が成り立つ。
③ 界面での反応平衡は(4-38)式のようになる。
④ 化学量論性を考慮したモル流束の関係式は(4-39)式で表される。

これらの式からC_o^*を消去して$C_c^*, C_c^b, k_c, k_o, K$などの諸量の関係式を導出し、続いて$C_c^* \simeq C_c^b$となる条件を求める。

(4-36)〜(4-39)式より、(4-40)式を得る。

3. Nonlinear thermodynamics

(4-38)式と(4-40)式からC_o^*を消去すると(4-41)式を得る。

これより、C_c^*を求めると(4-43)式のようになる。

この式から$C_c^*=C_c^b$となるような解を一般的に求めることは困難であるが、以下の手順に沿っておおよその値を求めることは可能である。

まず(4-43)式の右辺分子の第一項に着目し、①(4-44)式が成立する場合、さらに右辺分子の第二項に着目し、②(4-45)式が成立する場合、この①および②を同時に満たす条件において$C_c^*=C_c^b$となる。

先の条件[a]〜[d]の値を(4-44)式、および(4-45)式に代入し、これらを共に満たすC_c^bの最小値は0.095(wt%)であることがわかる。

すなわちC_c^bが0.095%より十分に大きい場合、炭素の移動は脱炭反応の律速とはならないと言える。

Substituting eq.(4-38) into eq.(4-40)

$$k_c(C_c^b - C_c^*) = k_o\left(C_c^b - \frac{P_{co}^*}{KC_c^*}\right) \quad (4\text{-}41)$$

$$\therefore Kk_c C_c^{*2} + (Kk_o C_o^b - Kk_c C_c^b)C_c^* - k_o P_{co}^* = 0 \quad (4\text{-}42)$$

$$\boxed{\therefore C_c^* = \frac{K(k_c C_c^b - k_o C_o^b) + \sqrt{(Kk_c C_c^b - k_o KC_o^b)^2 + 4k_o P_{co}^b Kk_c}}{2Kk_c}} \quad (4\text{-}43)$$

If transport of carbon to bubbule/metal interface is not rate controlling, $C_o^b \simeq C_c^*$.

For that to be valid,

then
$$\begin{cases} k_c C_c^b \gg k_o C_o^b & (4\text{-}44) \\ (Kk_c C_c^b)^2 \gg 4k_o P_{co}^b Kk_c & (4\text{-}45) \end{cases}$$

$$\rightarrow C_c^* = \frac{k_c C_c^b K}{2Kk_c} + \frac{\sqrt{(Kk_c C_c^b)^2}}{2Kk_c} = C_c^b \quad (4\text{-}46)$$

Applying the conditions [a] - [d] to eqs. (4-44) & (4-45),

$$C_c^b \gg \frac{k_o}{k_c} C_o^b = \frac{4\times 10^{-2}}{5.3\times 10^{-2}} \times 0.1 = 0.0755 \text{ wt\%}$$

$$(C_c^b)^2 \gg \frac{4k_o P_{co}^b}{k_c K} = \frac{4\times 4\times 10^{-2}\times 1.5}{5.3\times 10^{-2}\times 500} = 0.00905$$

$$\rightarrow C_c^b \gg 0.095 \text{wt\%}$$

If C_c^b drops to 0.095%, then transport of carbon to bubble interface may become rate controlling.

4. The chemical reaction which has an infinitely large equilibrium constant

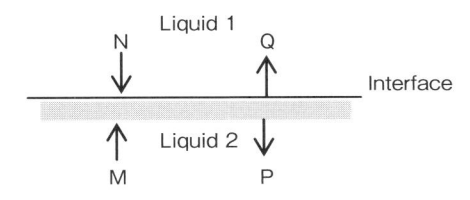

Fig. 4-6 Mass transfer and reaction at the interface of double liquid system (2).

きわめて大きな平衡定数の界面反応が関与する物質移動の律速課程の考え方についていくつかのパターンを概説する。

2液相界面において(4-47)式の反応が進行しており、その反応平衡定数が非常に大きな値を持つ((4-48)式)場合、界面におけるC_M^*、C_N^*の値のどちらか、または双方を0とみなすことができる((4-49)式)。

$$mM + nN \rightarrow pP + qQ \tag{4-47}$$

$$K = \frac{C_P^{*p} C_Q^{*q}}{C_M^{*m} \cdot C_N^{*n}} \rightarrow \infty \tag{4-48}$$

$$C_M^* \text{ and/or } C_N^* \rightarrow 0 \tag{4-49}$$

The implications of this statement are that for a reaction with a very large equilibrium constant the interfacial concentration of the rate controlling species is zero.

(1) Transport of M rate controlling

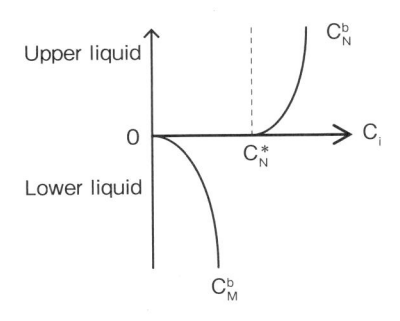

Fig. 4-7 Concentration profile in case of M rate control.

Mの移動が律速となる場合、界面近傍でのM、Nの濃度プロフィールを左図のように考えることができる。

Mを含有する液相の体積をV_b、2液相の界面積をAとすると、物質収支は(4-50)式、モル流束は(4-51)式、化学量論性の式は(4-52)式で表される。

①molar balance
$$\frac{\dot{n}_M}{A} A = -V_b \frac{dC_M^b}{dt} \tag{4-50}$$

②molar flux
$$\frac{\dot{n}_M}{A} = k_M(C_M^b - \cancel{C_M^*}^{\,0}) \tag{4-51}$$

③stoichiometry
$$\frac{\dot{n}_M}{A} = -\frac{m}{n}\frac{\dot{n}_N}{A} \tag{4-52}$$

(2) Transport of N rate controlling

Nの移動が律速となる場合

界面近傍でのM、Nの濃度プロフィールを右図のように考えることができる。

Nを含有する液相の体積をV_s、2液相の界面積をAとすると、物質収支は(4-53)式、モル流束は(4-54)式、化学量論性の式は(4-55)式で表される。

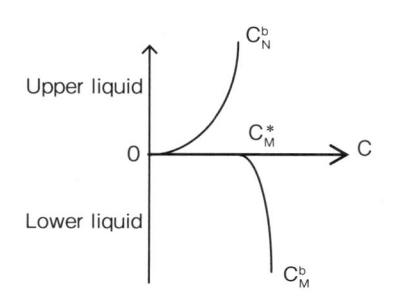

Fig. 4-8　Concentration profile in case of N rate controling.

①molar balance
$$\frac{\dot{n}_N}{A} A = V_s \frac{dC_N^b}{dt} \tag{4-53}$$

②molar flux
$$\frac{\dot{n}_N}{A} = -k_N(C_N^b), \text{ because } C_N^* = 0 \tag{4-54}$$

③stoichiometry
$$\frac{\dot{n}_M}{A} = -\frac{m}{n}\frac{\dot{n}_N}{A} \tag{4-55}$$

(3) Predicting which of the step is rate controlling

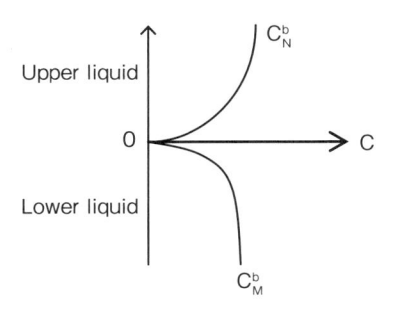

Fig. 4-9 Concentration profile in case of M or N rate controling.

C_M^*、C_N^*共に0の場合、M、Nどちらの移動が律速となるかを判断する。

① molar flux

$$\frac{\dot{n}_M}{A} = k_M(C_M^b - \cancel{C_M^*}^{0}) \quad (4\text{-}56)$$

$$-\frac{\dot{n}_N}{A} = k_N(C_N^b - \cancel{C_N^*}^{0}) \quad (4\text{-}57)$$

② stoichiometry

$$\frac{\dot{n}_M}{A} = -\frac{m}{n}\frac{\dot{n}_N}{A} \quad (4\text{-}58)$$

from eqs.(4-56)〜(4-58),

$$k_N C_M^b = \frac{m}{n} k_N C_N^b \quad (4\text{-}59)$$

※N is rate controlling

$$C_M^b \gg \frac{m}{n}\frac{k_N}{k_M} C_N^b \quad (4\text{-}60)$$

※M is rate controlling

$$C_M^b \ll \frac{m}{n}\frac{k_N}{k_M} C_N^b \quad (4\text{-}61)$$

M、Nについてのモル流束は(4-56)、(4-57)式で表される。
また、化学量論性の式は(4-58)式のようになる。(4-56)〜(4-58)式より(4-59)式を得る。
Nの移動が律速の場合、またMの移動が律速の場合、それぞれ(4-60)式、(4-61)式の関係が成り立つ事になる。

V. Application of Mass Transfer Analysis for Continuous Processing

1. Differences in general formulation for batch and continuous process

Application of ① molar flux expression, ② stoichiometry, and ③ chemical equilibrium to a relation yields the same equation whether the process is batch or continuous. The molar balance expression differs depending on the process.

連続プロセスにおける物質移動解析の適用について解説する。
①モル流速、②化学量論性、③化学平衡については、プロセスがバッチ、連続にかかわりなく同等の式が成り立つ。
物質収支はプロセスによって異なる。

(1) Batch process

Choose a control volume and assume some fluid flow or mixing conditions. For example, assuming back mixed reactor (perfectly mixed),

$$\text{Rate of i input} \quad \frac{(\dot{n}_i)_{in}}{A} A \tag{5-1}$$

$$\text{Rate of i output} \quad \frac{(\dot{n}_i)_{out}}{A} A \tag{5-2}$$

$$\text{Rate of generation} \quad (\dot{r}_i) V \tag{5-3}$$

$$\text{Rate of accumulation} \quad V \frac{dC_i^b}{dt} \tag{5-4}$$

$$\therefore \boxed{(\dot{n}_i)_{in} + \dot{r}_i V = (\dot{n}_i)_{out} + V \frac{dC_i^b}{dt}} \tag{5-5}$$

たとえば、完全混合の反応槽を考えると、界面（面積A）を通しての成分iのインプット速度は(5-1)式、アウトプット速度は(5-2)式、反応槽内でのiの生成速度は反応速度を$\dot{r}_i$、槽の体積をVとすると(5-3)式、iの蓄積速度は(5-4)式となり、(5-5)式で示すバランスが成立する。

(2) Continuous process

$$\text{Rate of input of i at AC} = C_i L 1 u \tag{5-6}$$

$$\text{Rate of output of i at BD} = u1LC_i + \frac{dC_i}{dx} dx u1L \tag{5-7}$$

$$\text{Rate of output of i at AB} = \frac{n_i}{A} dx 1 \tag{5-8}$$

$$\text{Rate of accumulation of i in ABCD} = 0 \tag{5-9}$$

図5-1に示すような連続プロセスを想定する。
上部と下部（高さL）の液相または気相は両方共y方向に完全混合されており、下部の液相はx方向にプラグフロー（押し出し流れ）となっている。
なお、以下の定式化においては相の奥

行き（厚み）を1として行う。

ABCDの体積要素（幅dx、高さL）において、AC面から入るiの速度（単位はmole/s）は(5-6)式で、BD面から出るiの速度は(5-7)式で与えられる。

また、AB面から上部液相に出るiの速度は(5-8)式となる。

なお、このプロセスは定常状態にあるため、iの蓄積は無い。

これらから物質収支は(5-10)式のようになり、最終的に(5-11)式となる。

ここでモル流束（$\frac{\dot{n}_i}{A}$）はiの濃度（C_i）で表される。

また、バッチプロセスと同様に、連続プロセスにおいても反応界面においては平衡の式や化学量論性の式が成り立っている。

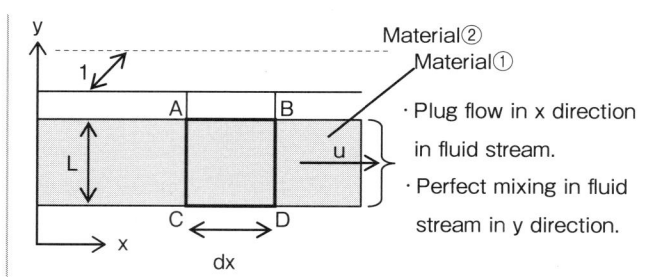

Fig. 5-1 Schematic figure of continous refining Process.

- Plug flow in x direction in fluid stream.
- Perfect mixing in fluid stream in y direction.

Since the process is steady state,

$$uLC_i - uLC_i - uL\frac{dC_i}{dx}1dx - \frac{\dot{n}_i}{A}1dx = 0 \quad (5\text{-}10)$$

$$\therefore \boxed{uL\frac{dC_i}{dx} + \frac{\dot{n}_i}{A} = 0} \quad (5\text{-}11)$$

$\frac{\dot{n}_i}{A}$ can be expressed in terms of C_i using molar flux expressions.

Chemical equlibrium & stoichiometry of a chemical reaction occurs at the interface.

2. Case study (1); Deoxidation of molten blister copper using a submerged CO jet

Prior to casting copper from the converters the blister copper is put into an anode furnace, a furnace that refine the blisten copper to the anode-grade copper, in which the oxygen content must be reduced from approximately 1 to 0.1 wt%. The deoxidation, which traditionally had been effected by "poling", is now commonly accomplished by blowing a reducing gas such as methane or reform gas (CO or natural gas) through the bath from a submerged tuyere (vide Figs. 5-2 and 5-4).

溶融している粗銅の底吹きCOガスジェットによる脱酸反応について解析を行う。

銅の電気精錬で使用する「アノード」板の鋳造の前に、精錬炉において溶融酸素濃度を約1％から0.1％程度まで低下させる必要がある。

銅精錬の歴史をたどると「ポーリング」と呼ばれる「木材（ポール）の溶融粗銅中への投入」工程が精錬の由来となっている。新鮮な木材に含まれる樹液が蒸発して還元剤として作用することを利用したものである。現在ではCOガスや天然ガスなどの還元性ガスが炉底に設けられた羽口から吹き込まれている。

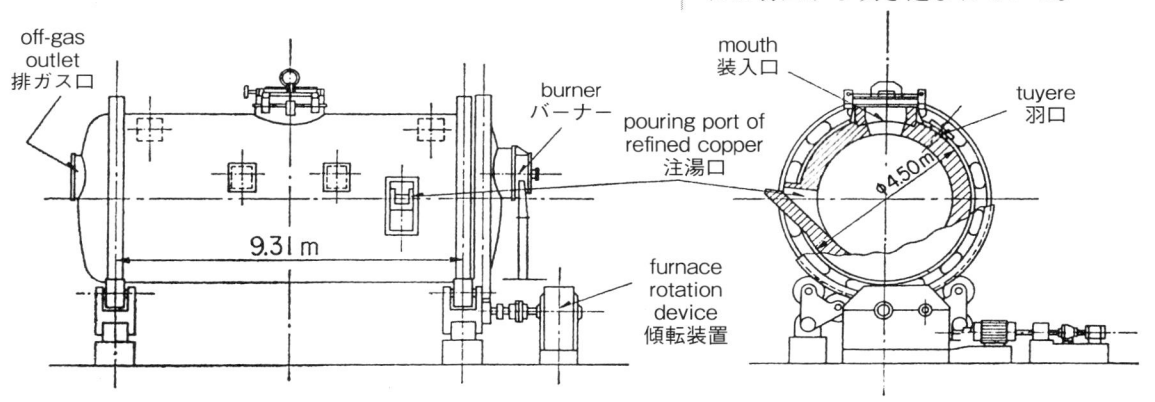

Fig. 5-2 Anode furnece for deoxidation of molten blister copper.[6]

The gas reacts with the oxygen dissolved in the copper to form CO_2 and H_2O which are swept out of the bath by the jet.

In order to identify the rate controlling steps and determine reaction rates in this process, small scale experiments have been conducted in which a jet of CO gas was blown through a Cu-O bath. The reaction between CO and dissolved oxygen ($CO + O_{Cu} = CO_2$) has a very large equilibrium constant and is known to be virtually instantaneous at the test temperature of 1170°C.

For a constant jet Reynolds number the oxygen

還元性ガスは溶融銅中の酸素と反応してCO_2やH_2Oとなるが、これらは吹き込まれるガスと共に系外に排出される。

このプロセスにおける反応速度や律速段階について調査を行うために、酸素を含む銅中にCOガスを吹き込む小規模実験が行われた。

COと銅中酸素の反応は非常に大きい平衡定数を有する事が知られている。

所定のジェットレイノルズ数において銅中の酸素濃度自体が銅の還元速度に影響することが分かっている（図5-3）。

concentration in the copper was found to affect the reaction rate according to Fig. 5-3.

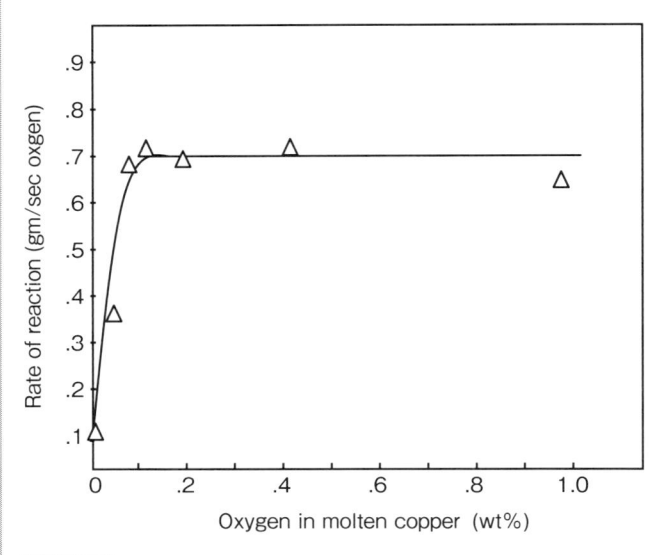

Fig. 5-3 Effect of oxygen concentration in copper on the rate of reaction.

(1) Consider the rate limiting steps in the reaction and explain why the reaction rate is dependent on oxygen concentrations at low concentrations but is independent of oxygen concentration at higher concentrations. With the following information at hand calculate the value of oxygen concentration intermediate between the two types of behaviour. Ignore any effect of the cuprostatic head.

Gas constant = 82 cm^3 atm mole^{-1}K^{-1}

Gas phase mass transfer number ($k_{co}a'$)
 = 150 cm^2 sec^{-1}

Liquid phase mass transfer number ($k_o a'$)
 = 2.60 cm^2 sec^{-1}

Density of copper at 1170℃ = 7.9 g cm^{-3}

(0.1 wt% of orygen in copper
 = 4.94 (10^{-4}) mole cm^{-3})

〈Note〉
The mass transfer number is the product of the mass

この反応における律速段階について考察すると共に、低酸素濃度の領域で反応速度は酸素濃度の影響を受ける一方、高酸素濃度になると酸素濃度に依存しなくなるという傾向を、提示された条件の下に説明する。溶融銅の深さによる圧力変化は無視する。

物質移動数とは、物質移動係数とジェ

transfer coefficient and the interfacial area per unit length of the jet, α'.

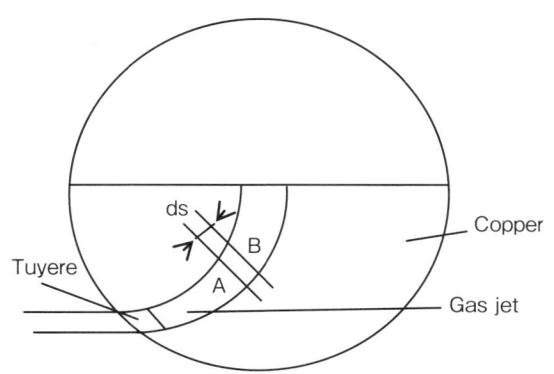

Fig. 5-4 Jet injected into a molten copper bath.

α' : surface area of jet per unit length of jet trajectory
s : distance along jet trajectory

図5-4に示すように、ジェットにおいて面AとBに挟まれた厚みdsの体積要素を考える。
この体積要素における物質収支を取る。
面Aから入るiの速度は(5-12)式、面Bから出るiの速度は(5-13)式で示される。
また、ジェットと溶融銅の界面から出るiの速度は(5-14)式となる。ここでAは界面積である。

(5-12)～(5-14)式より、物質収支の式は(5-15)式となる。

Apply a molar balance to a control volume if Q=volume flow rate of gas at inlet of tuyer, P_i=partial pressre of i component.

$$\text{Rate of input of i at A} = \frac{P_i}{RT} Q \quad \frac{\text{mole}}{s} \tag{5-12}$$

$$\text{Rate of out put of i at B} = \frac{P_i}{RT} Q + \frac{Q}{RT} \frac{dP_i}{ds} ds \tag{5-13}$$

$$\text{Rate of out put of i at jet/metal interface} = \frac{\dot{n}_i}{A} \alpha' ds \tag{5-14}$$

Then, equation of mass balance is,

$$\boxed{-\frac{Q}{RT} \frac{dP_i}{ds} ds - \frac{\dot{n}_i}{A} \alpha' ds = 0} \tag{5-15}$$

$$CO_{(g)} + [O]_{cu} \rightarrow CO_{2(g)} \tag{5-16}$$

(i) Since the chemical reaction is instantaneous, the reaction is transport controlled.
(ii) Since the reaction has a large equilibrium

続いて、ジェット／溶融銅間における脱酸反応((5-16)式)に関しては(i)、(ii)とみなすことが出来るため、(5-17)式、(5-18)式が成り立つ。

constant, the transport of $[O]_{Cu}$ or CO in the jet must be rate limiting.

$$K = \frac{P^*_{co_2}}{P^*_{co} C^*_o} \to \infty \quad (5\text{-}17)$$

$$P^*_{co} \to 0 \text{ and/or } C^*_o \to 0 \quad (5\text{-}18)$$

<a> If transport of oxygen is rate limited,

$$\frac{\dot{n}_o}{L} = k_o \alpha'(C^b_o - C^*_o) \xrightarrow{C^*_o \to 0} \quad (5\text{-}19)$$

溶銅中の酸素の移動が律速の場合、モル流束は(5-20)式となる。ここでLはジェット長さである。

$$= k_o \alpha'(C^b_o) = \frac{\dot{n}_{co}}{L} \quad (5\text{-}20)$$

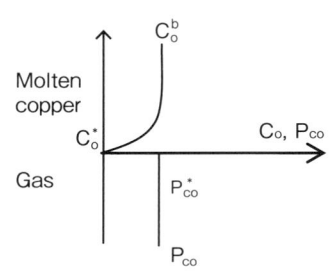

Fig. 5-5 Concentration profile near the interface when transport of oxygen is rate limited.

一方、ガス中のCOの移動が律速の場合、モル流束は(5-22)式のようになる。

<b> If transport of CO is rate limited,

$$\frac{\dot{n}_{co}}{L} = \frac{k_{co} \alpha'}{RT}(P^b_{co} - P^*_{co}) \xrightarrow{P^*_{co} \to 0} \quad (5\text{-}21)$$

$$\frac{\dot{n}_{co}}{L} = \frac{k_{co} \alpha'}{RT}(P^b_{co}) = \frac{\dot{n}_o}{L} \quad (5\text{-}22)$$

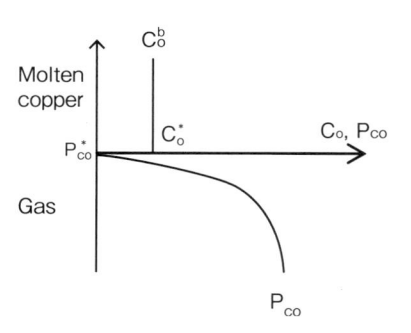

Fig. 5-6 Concentration profile near the interface when transport of CO is rate limited.

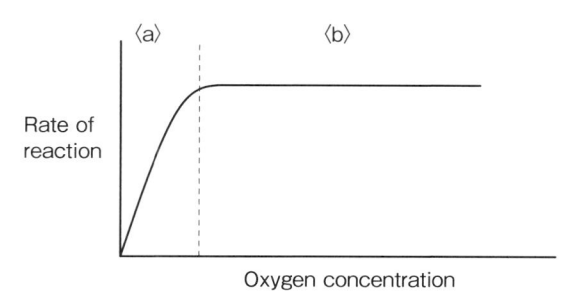

Fig. 5-7 Change of reaction rate with the concentration of oxygen in copper.

In region ＜a＞ equation (5-20) holds & rate of reaction is a linear function of oxygen concentrations.

In region ＜b＞ equation (5-22) holds & rate of reaction is independent of oxygen concentrations.

At transition from ＜a＞ to ＜b＞, both equations are valid.

$$k_o\alpha'(C_o^b) = \frac{k_{co}\alpha'}{RT}(P_{co}^b) \qquad (5\text{-}23)$$

$$\underset{\text{transition}}{C_{o(t)}^b} = \frac{k_{co}\alpha'}{k_o\alpha'}\frac{P_{co}^b}{RT} = \frac{150}{2.60}\frac{1}{82\cdot 1443}$$

$$= 4.88\times 10^{-4}\,\text{moles/cm}^3$$

$$= 0.099\,\text{wt\%} \qquad (5\text{-}24)$$

すなわち、領域＜a＞においては(5-20)式が成り立ち、反応速度は溶銅バルク中の酸素濃度と一次の関数となり、領域＜b＞においては(5-22)式が成り立って、反応速度は酸素濃度の影響を受けない事がわかる。

さらに、〈a〉と〈b〉の領域の境界においては(5-23)式が成立するため、領域遷移における酸素濃度$C_{o(t)}^b$は(5-24)式のように表され、その値は図5-3と一致する。

(2) The mass transfer numbers obtained in these experiments have been scaled up to the operating conditions of anode copper deoxidation at W Company where CO is injected through a tuyere at 184,000 cm³ sec⁻¹. The $k_{co}\alpha'$ has a value of 96,000 cm² sec⁻¹. If the length of the jet trajectory is 95cm, calculate the partial pressure of CO in the jet at the surface of the bath. Assume that the volumetric flow rate of gas in the jet is constant.

If we apply a molar balance to the jet element "ds",

Rate of input of CO by bulk motion $= \dfrac{QP_{co}^b}{RT}$ (5-25)

この実験で得られた物質移動数を実機の操業条件にスケールアップする場合、その値は96000cm²sec⁻¹となる。その際の実機におけるプール表面でのジェット中のCO分圧を以下に推算する。

(1)と同様に定常状態においては(5-28)式が成立する。
さらに、ジェット内は半径方向に完全

混合で、軸方向（ジェットの向かう方向）には押し出し流れであると仮定する。

酸素が高い段階（反応速度が酸素濃度に依存しない領域）において、COガス生成のモル流束は(5-29)式で表される。

Rate of output of CO by bulk motion=
$$\frac{QP_{co}^b}{RT} + \frac{d}{ds}\left(\frac{QP_{co}^b}{RT}\right)ds \quad (5\text{-}26)$$

Rate of ortput of CO at jet/metal interface $= \frac{\dot{n}_{co}}{L}ds \quad (5\text{-}27)$

At the steady state,
$$-\frac{Q}{RT}\frac{dP_{co}^b}{ds} - \frac{\dot{n}_{co}}{L} = 0 \quad (5\text{-}28)$$

We have assumed for the jet
(a) back mixing radially
(b) plug flow axially

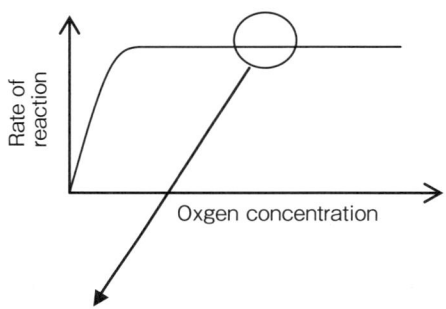

in this stage

$$\frac{\dot{n}_{co}}{L} = \frac{k_{co}\alpha'}{RT}(P_{co}^b) \quad (5\text{-}29)$$

$$\therefore -\frac{Q}{RT}\frac{dP_{co}^b}{ds} - \frac{k_{co}\alpha'}{RT}P_{co}^b = 0 \quad (5\text{-}30)$$

(5-30)式を、(5-31)式に示すようにP_{co}^bについて羽口での分圧(P_{co}^{bi})から液面での分圧(P_{co}^{bo})まで、Sについて羽口(0)から液表面(S_0)まで積分すると、(5-32)式が得られる。

これより、液面でのCO分圧は2.98×10^{-22}atmと求まる。

$$\int_{P_{co}^{bi}}^{P_{co}^{bo}} \frac{dP_{co}^b}{P_{co}^b} = -\frac{k_{co}\alpha'}{Q}\int_0^{S_o} ds \quad (5\text{-}31)$$

$$\ln\left(\frac{P_{co}^{bo}}{P_{co}^{bi}}\right) = \frac{-k_{co}\alpha' S_o}{Q} \quad (5\text{-}32)$$

$$P_{co}^{bo} = (1\text{atm})\exp\left\{-\frac{(96{,}000\text{cm}^2/\text{s})(95\text{cm})}{(184{,}000\text{cm}^3/\text{s})}\right\}$$
$$= 2.98 \times 10^{-22}\text{atm} \quad (5\text{-}33)$$

溶銅中の酸素濃度が0.1%近くに減少すると、羽口では溶融銅バルクから銅／ジェット界面へ酸素の移動が、それ以上では銅／ジェット界面からジェットバルクへのCOの移動が反応を律速する状況

(3) As the oxygen concentration in the bath drops to 0.1 wt%, transport control at the tuyere exit changes to transfer of oxygen from the melt to the jet interface, while

in the rest of the jet the reaction rate is controlled by mass transfer of CO from the bulk of the jet to the interface.

As the C_o^b in the bath drops further, the lower part of jet becomes O-transport control while the upper part of the jet remains CO transport control. This phenomenon is called "consecutive control". Prove that for consecutive control to be possible when the process conditions are expressed as eq. (5-34)[8].

$$\boxed{\frac{1}{1+\frac{k_{co}\alpha'S_o}{Q}} < \frac{k_o\alpha'C_o^b RT}{k_{co}\alpha'P_{co}^b} < 1} \qquad (5-34)$$

になる。この現象は"consective control"(遷移律速)と呼ばれる。この状態にある時、プロセス変数および定数は(5-34)式の関係となることを証明する。

in this stage $\dfrac{\dot{n}_{co}}{L} = \dfrac{k_{co}\alpha'}{RT}P_{co}^b$ (i)

in this stage $\dfrac{\dot{n}_{co}}{L} = k_o\alpha'C_o^b$ (ii)

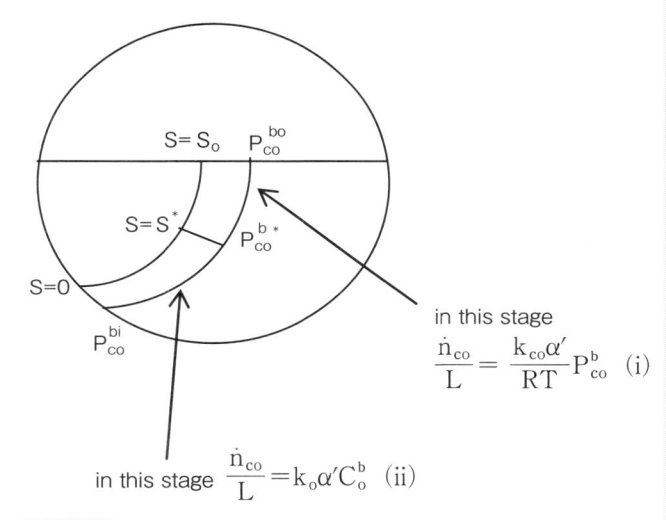

Fig. 5-8 Consective control for a deoxydation of copper.

$$\frac{dP_{co}}{ds}\frac{Q}{RT} + \frac{\dot{n}_{co}}{L} = 0 \qquad (5\text{-}35)$$

$$\frac{dP_{co}}{ds}\frac{Q}{RT} + k_o\alpha'C_o^b = 0 \qquad (5\text{-}36)$$

$$\int_{P_{co}^{bi}}^{P_{co}^{b*}} dP_i = -\frac{C_o^b k_o\alpha'RT}{Q}\int_0^{s*} ds \qquad (5\text{-}37)$$

$$\therefore \boxed{P_{co}^{b*} - P_{co}^{bi} = -\frac{C_o^b k_o\alpha'RT}{Q}S^*} \qquad (5\text{-}38)$$

from eqs. (i) & (ii) in Fig 5-8

図5-8に示すようにS=S*を境にして下部はメタルからジェットとの界面へのOの移動が反応を律速し、上部はジェットバルクからメタルとの界面へのCOの移動が反応を律速することになる。

下部では(5-36)式が成り立つがこれをP_{co}, sについてそれぞれP_{co}^{bi}からP_{co}^{b*}、0からS*まで積分すると(5-38)式が得られる。

図5-8中の式より(5-39)式を得て、

(5-38)式と組み合わせてS^*は(5-42)式のように表される。

S^*は(5-43)式の領域にあるため、(5-44)式となり、これより "consective control" の条件を示す(5-47)式を得る。

$$P_{co}^{b*} = \frac{RTk_oC_o^b}{k_{co}} \tag{5-39}$$

from eqs. (5-38) & (5-39)

$$\frac{RTk_oC_o^b}{k_{co}} - P_{co}^{b_i} = -\frac{C_o^b k_o \alpha' RT}{Q} S^* \tag{5-40}$$

$$\therefore S^* = \left(P_{co}^{b_i} - \frac{RTk_oC_o^b}{k_{co}}\right)\frac{Q}{C_o^b k_o \alpha' RT} \tag{5-41}$$

$$\boxed{S^* = \left(\frac{P_{co}^{b_i}}{C_o^b k_o \alpha' RT} - \frac{1}{k_{co}\alpha'}\right)Q} \tag{5-42}$$

$$\boxed{0 < S^* < S_o} \tag{5-43}$$

$$\therefore 0 < \left(\frac{P_{co}^{b_i}}{C_o^b k_o \alpha' RT} - \frac{1}{k_{co}\alpha'}\right)Q < S_o \tag{5-44}$$

$$\therefore \frac{Q}{k_{co}\alpha'} < \frac{P_{co}^{b_i}Q}{C_o^b k_o \alpha' RT} < S_o + \frac{Q}{k_o\alpha'} \tag{5-45}$$

$$\therefore 1 < \frac{P_{co}^{b_i}k_{co}\alpha'}{C_o^b k_o \alpha' RT} < \frac{S_o k_{co}\alpha'}{Q} + 1 \tag{5-46}$$

$$\boxed{\therefore \frac{1}{1 + \frac{S_o k_{co}\alpha'}{Q}} < \frac{C_o^b RT k_o \alpha'}{P_{co}^{b_i}\alpha' k_{co}} < 1} \tag{5-47}$$

<Note>

In the first stage of refining copper, sulfur and iron are removed by gently blowing air through the molten metal to form iron oxides and sulfur dioxide. The second stage involves using a reducing agent, normally natural gas, is used to react with the oxygen in the copper. In the past, freshly cut trees were used as wooden poles. The sap in these poles acted as the reducing agent. It was the use of these poles gave rise to the term "poling".

3. Case study (2); Kinetics of vacuum dezincing of lead

3. 1 Process description

The continuous vacuum dezincing process has been studied some time ago as a method of separating zinc from molten lead in the Imperial Smelting process.[9)]

In the vacuum dezincing process, lead containing initially about 2.2% of zinc is directed to flow down a spiral launder which is welded to the inside wall of chamber as shown in Fig.5-9. During its passage down the launder, zinc evaporates from the surface of the molten lead and condenses on a concentrically positioned water cooled pipe. The liquid lead and zinc are removed via barometric legs at the bottom.

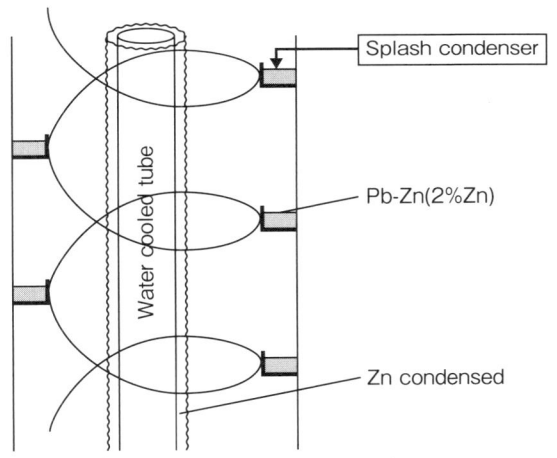

第1章で述べたようにImperial Smelting炉では亜鉛蒸気を含むガスが炉頂から取り出される。このガスは溶融鉛の細粒が噴霧されるスプラッシュコンデンサーを通過する際に溶融鉛に吸収される。亜鉛を含む溶融鉛は冷却されて、亜鉛の一部はコンデンサー出側で冷却分離される。分離後の約2%の亜鉛を含む溶融鉛はコンデンサーに送り返され、再び亜鉛の吸収に使用されるという循環系を構成している。

ここでは、この亜鉛を含む溶融鉛から亜鉛を抽出する"連続真空亜鉛分離プロセスにおける亜鉛抽出効率見積りの理論検討の一部"をケーススタディとして紹介する。

図5-9に示すように真空炉の内壁に螺旋状に樋が設置されている。この螺旋状樋を流れる亜鉛を2.2%程度含む溶融鉛の自由表面で亜鉛の蒸発は進行する。蒸発した亜鉛は真空炉の中央に設置された内部を水冷しているシリンダー表面で凝縮される。

この炉内の真空度は約0.1mmHgであり、処理を終えた溶融鉛と抽出した溶融亜鉛は大気脚(barometric leg)によって系外に取り出され、溶融鉛は再び凝縮器に戻される。

Fig. 5-9 Continuous vacuum dezincing system in the Imperial Smelting process.

3. 2 Modeling of the continuous vacuum zinc distillation

We have to determine the concentration of zinc in lead, C_{Zn}, as a function of position x along a length of the spiral launder.

このプロセスを設計するに当たって、まず真空炉による液体鉛からの亜鉛の抽出効率を推定する必要があるが、そのために、樋の長さ方向の溶融鉛中の亜鉛の濃度変化を推定するモデルを検討する。

a) 系は定常状態であり溶融鉛中の亜鉛濃度、C_{Zn}^bは樋の長さの関数である。
b) 樋中の流れはプラグフロー（押し出し流れ）で速度は一定である。
c) 樋中の溶融鉛の亜鉛濃度は幅方向に一定である。
d) 樋中の溶融鉛の温度は一定である。

図5-10に樋中の亜鉛を含む溶融鉛の流れを示す。流れの方向をx、幅をw、深さをℓ、速度をuとし、その中の要素体積（長さdx）を考える。
① 流れから要素体積中に入る亜鉛の速度は $uw\ell C_{Zn}^b$
② 要素体積から出る亜鉛の速度は
$uw\ell C_{Zn}^b + \dfrac{d}{dx}(uC_{Zn}^b \ell w)dx$
③ 要素体積の表面から真空中に移動する亜鉛の速度は、定常状態では溶融鉛のバルクから界面に移動する亜鉛の速度と同等と見なすことができるため
$\left(\dfrac{\dot{n}_{Zn}}{A_e}\right)w\,dx$と表す。ここで面積は蒸発面積Aeを使用している。流れのある要素体積では、その表面積と等しいとみなすことができる。

定常状態におけるこれらの収支を取ると、(5-51)式のようになり、亜鉛のフラックスは(5-52)式となる。

〈Assumptions〉

a) Steady state operation; $C_{Zn}^b = f(x)$
b) Plug flow in the direction of motion.
c) Negligible gradient of C_{Zn}^b in the width of the launder.
d) Constant flow velocity and constant temperature.

w : width of flow in launder
ℓ : depth of flow in launder
x : flow direction
u : flow velocity

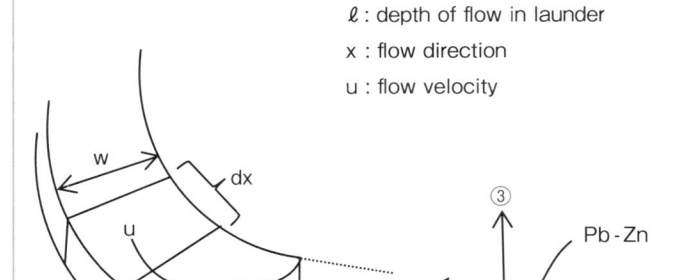

Fig. 5-10 Continuous flow of liquid Pb-Zn in the launder equipped inside the vacuum chamber.

〈Equations〉

① Rate of zinc input to the control volume

$$uw\ell C_{Zn}^b \qquad (5\text{-}48)$$

② Rate of zinc output from the control volume

$$uw\ell C_{Zn}^b + \dfrac{d}{dx}(uC_{Zn}^b \ell w)dx \qquad (5\text{-}49)$$

③ Rate of zinc output by evaporation

$$\left(\dfrac{\dot{n}_{Zn}}{A_e}\right)w\,dx \qquad (5\text{-}50)$$

$\dot{n}_{Zn}$: rate of zinc evaporation

A_e : area of evaporation

$$\therefore -uw\ell \dfrac{d}{dx}C_{Zn}dx - \dfrac{\dot{n}_{Zn}}{A_e}w\,dx = 0 \qquad (5\text{-}51)$$

$$\therefore -u\ell \dfrac{d}{dx}C_{Zn} - \dfrac{\dot{n}_{Zn}}{A_e} = 0 \qquad (5\text{-}52)$$

The next step is to estimate the rate of zinc distillation as a function of process variables.

ⓐ Transport of Zn from bulk to surface.

$$\frac{\dot{n}_{Zn}}{A_e} = k_{Zn}(C_{Zn}^b - C_{Zn}^*) \quad (5\text{-}53)$$

k_{Zn}: a liquid-phase mass transfer coefficient

ⓑ Evaporation of Zn from surface.

$$E = \alpha_e A_e\, a\{P^* - P_1(1-r_1)^2\} \quad (5\text{-}54)$$

α_e : accommodation coefficient, $0 < \alpha_e < 1$

A_e : evaporation area

$a\ \ :\ \dfrac{1}{(2\pi MRT)^{\frac{1}{2}}}$

M : molecular weight

P^* : vapour pressure of Zn at given temperature T

P_1 : pressure immediately adjacent to the evaporation surface

r_1 : coefficient which accounts for a reduction in a static pressure due to kinetic energy of leaving Zn molecule

T : evaporation temperature

ⓒ Condensation of Zn

$$C = \alpha_c A_c(a(1+r_2)^2 P_2 - bP_c) \quad (5\text{-}55)$$

α_c: accommodation coefficient, $0 < \alpha_c < 1$

A_c : condensation area

$b\ \ :\ \left(\dfrac{1}{2\pi MRT_c}\right)^{\frac{1}{2}}$

r_2 : coefficient which accounts for an increase in a static pressure due to kinetic energy of reaching Zn molecule

続いて、このプロセスにおける亜鉛の蒸留速度をプロセス変数の関数として求める。

ⓐ溶融鉛の要素体積バルクから真空界面への移動速度 $\dot{n}_{Zn}$、ⓑ要素体積表面からの亜鉛の蒸発速度 E、ⓒ亜鉛の凝集速度 Cについては左記のように記述される。

ⓐの(5-53)式において、k_{Zn}は溶融鉛／真空界面での亜鉛の物質移動係数、ⓑの(5-54)式において、α_eは蒸発における適応係数、A_eは蒸発面積、aはラングミュア項 $\dfrac{1}{(2\pi MRT)^{\frac{1}{2}}}$（Mは分子量、Rは気体定数、Tは蒸発温度）。$P^*$は温度Tにおける亜鉛の蒸気圧、$P_1$は蒸発面直上の圧力、$r_1$は蒸発表面から去る亜鉛分子の運動エネルギーによる静圧の減少を補正する係数である。

ⓒの(5-55)式において、α_cは凝集における適応係数、A_cは凝集面積、bはラングミュア項 $\dfrac{1}{(2\pi MRT_c)^{\frac{1}{2}}}$（$T_c$は凝集温度）、$r_2$は凝集面に入る亜鉛分子の運動エネルギーによる静圧の減少を補正する係数、P_2は凝集面直上の圧力、P_cは凝集温度における平衡圧力である。

P_2 : pressure immediately adjacent to the condensation surface

P_c : equilibrium vapor pressure at condenser temperature T_c

定常状態においては(5-56)式のように $\dot{n}_{Zn} = E = C$ が成り立つため、これと(5-53)～(5-55)式を組み合わせて圧力項および界面濃度や界面圧力の項が含まれない式を右記のように導出する。

Under steady state conditions of no build-up or depletion at the interface the mass transfer rate is equal to the distillation rate, and by combining equations (5-53), (5-54) and (5-55) the theoretical rate equation is derived as follows;

$$\dot{n}_{Zn} = E = C \qquad (5\text{-}56)$$

Provided the concentration of zinc at the interface C^*_{Zn} is in equilibrium with the partial pressure P^*,

ここで界面で濃度 C^*_{Zn} と圧力 P^* は平衡関係にありこの比をKとすると((5-57)式)、液相中の物質移動の式(5-53)式と組み合わせて(5-58)式が得られる。

$$K = \frac{P^*}{C^*_{Zn}} \qquad (5\text{-}57)$$

Note that K contains the product of the vapour pressure of pure solute at the evaporation temperature, activity coefficient of the solute referred to its concentration at the surface and numerical factor to keep units consistent.

from eqs. (5-53) and (5-57)

$$\dot{n}_{Zn} \cdot \frac{K}{k_{Zn} A_e} = K C^b_{Zn} - K C^*_{Zn}$$
$$= K C^b_{Zn} - P^* \qquad (5\text{-}58)$$

(5-54)式を変形して(5-59)式を得るが、(5-58)式と(5-59)式を足し合わせることで(5-60)式となり、また $\dot{n}_{Zn} = E$ であるため(5-61)式となる。
凝集に関しては(5-55)式と(5-56)式より(5-62)式を得る。

from eq. (5-54)

$$\frac{E}{\alpha_e A_{ea}} = P^* - P_1(1-r_1)^2 \qquad (5\text{-}59)$$

from eqs. (5-58) and (5-59)

$$\therefore \frac{\dot{n}_{Zn} K}{k_{Zn} A_e} + \frac{E}{\alpha_e A_{ea}} = K C^b_{Zn} - P_1(1-r_1)^2 \qquad (5\text{-}60)$$

from eq. (5-56)

$$\dot{n}_{Zn}\left(\frac{K}{k_{Zn}A_e} + \frac{1}{\alpha_e A_e a}\right)\frac{1}{(1-r_1)^2} = \frac{KC_{Zn}^b}{(1-r_1)^2} - P_1 \quad (5\text{-}61)$$

from eq. (5-55)

$$\frac{\dot{n}_{Zn}}{\alpha_c A_c}\frac{1}{a(1+r_2)^2} = P_2 - \frac{bP_c}{a(1+r_2)^2} \quad (5\text{-}62)$$

As described in [Ref.],

$$P_1 = P_2 = P \quad (5\text{-}63)$$

from eqs. (5-61), (5-62) and (5-63)

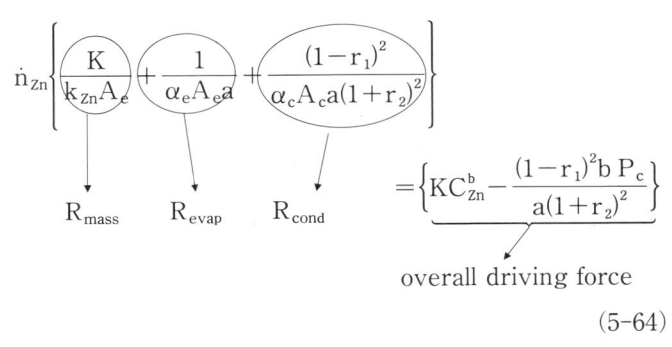

$$\dot{n}_{Zn}\left\{\underbrace{\frac{K}{k_{Zn}A_e}}_{R_{mass}} + \underbrace{\frac{1}{\alpha_e A_e a}}_{R_{evap}} + \underbrace{\frac{(1-r_1)^2}{\alpha_c A_c a(1+r_2)^2}}_{R_{cond}}\right\} = \underbrace{\left\{KC_{Zn}^b - \frac{(1-r_1)^2 bP_c}{a(1+r_2)^2}\right\}}_{\text{overall driving force}} \quad (5\text{-}64)$$

[Ref.] に示すように、真空中での分子の移動速度に大きな変化はなく $P_1 = P_2 = P$ と置くと(5-61)式、(5-62)式、(5-63)式より(5-64)式が得られる。(5-64)式の左辺{ }内は物質移動の全抵抗、右辺は駆動力をそれぞれ表す。

Finally the rate of zinc distillation is expressed as a function of process variables (see eq. (5-65)).

最終的に亜鉛の蒸留速度は(5-65)式のように表される。ここで右辺{ }の係数が総括物質移動抵抗である。

$$\dot{n}_{Zn} = \frac{k_{Zn}\alpha_c A_c A_e \alpha_e}{\alpha_c A_c(1+r_2)^2(\alpha_e aK + k_{Zn}) + k_{Zn}A_e\alpha_e(1-r_1)^2}\{aKC_{Zn}^b(1+r_2)^2 - (1-r_1)^2 bP_c\} \quad (5\text{-}65)$$

Identify the condition which each of the steps is rate controlling.

さて、以下にそれぞれのステップが律速となる場合について考えてみる。
case I は液相中物質移動が、case II は蒸発が、また case III は凝集が律速過程になる場合の関係である。

[case I] liquid phase mass transfer rate controlling

$$\frac{K}{k_{Zn}A_e} \gg \frac{1}{\alpha_e A_e a} + \frac{(1-r_1)^2}{\alpha_c A_c(1+r_2)^2} \quad (5\text{-}66)$$

[case II] evaporation resistance rate controlling

$$\frac{1}{\alpha_e A_e a} \gg \frac{K}{k_{Zn}A_e} + \frac{(1-r_1)^2}{\alpha_c A_c(1+r_2)^2} \quad (5\text{-}67)$$

続いて、蒸発の抵抗が他のステップに比べ小さい場合の亜鉛の移動速度を考えてみる。更なる簡略化のために、蒸発時や凝集時の亜鉛分子の移動に伴う圧力変化が無視でき（$r_1 = r_2 = 0$）、系の温度は一定（$T = T_1$）とすると(5-69)式が得られる。

最後に、(5-52)式のZnバランスに(5-69)式で得られた $\dot{n}_{Zn}/A_e$ を代入すると、目的とする樋長さ方向の亜鉛濃度の変化の式、(5-71)～(5-73)式を得ることができる。ここで(5-71)式を溶融鉛中の亜鉛初期濃度 C_{Zn}^i (at $x=0$) から C_{Zn}^f (at $x=L$) まで積分し、目的とする亜鉛析出量を得るための樋の長さLを求めることができる。

case Ⅲ condensation rate controlling

$$\frac{(1-r_1)^2}{\alpha_c A_c a(1+r_2)^2} \gg \frac{1}{\alpha_e A_e a} + \frac{K}{k_{Zn} A_e} \quad (5\text{-}68)$$

Calculate the length of the launder if evapolation resistance is small compared to the other two, $r_1 = r_2 = 0$, and $T = T_c$.

$$\dot{n}_{Zn} = \frac{1}{\left(\dfrac{K}{k_{Zn} A_e} + \dfrac{1}{\alpha_c A_c a}\right)} (KC_{Zn}^b - P_c) \quad (5\text{-}69)$$

We can now substitute this ($\dot{n}$) into zinc balance, eq(5-52).

$$-u\ell \frac{dC_{Zn}}{dx} - \left(\frac{1}{\dfrac{K}{k_{Zn} A_e} + \dfrac{1}{\alpha_c A_c a}}\right) \frac{(KC_{Zn}^b - P_c)}{A_e} = 0 \quad (5\text{-}70)$$

$$\frac{dC_{Zn}}{dx} = -\frac{R}{Q}(KC_{Zn}^b - P_c) \quad (5\text{-}71)$$

where

$$R = \frac{1}{\dfrac{K}{k_{Zn} A_e} + \dfrac{1}{\alpha_c A_c a}} \quad (5\text{-}72)$$

$$Q = A_e u \ell \quad \sim \text{flow rate} \quad (5\text{-}73)$$

The length of the launder, L, can be obtained by integrating eq.(5-71) with the following conditions,

$x = 0$; $C_{Zn} = C_{Zn}^i$ (initial zinc concentration in launder)

$x = L$; $C_{Zn} = C_{Zn}^f$ (final zinc concentration in launder)

[Ref.] Rates of evaporation, condensation and liquid-phase mass Transfer[9)]

The transport of the distilling species across the distillation gap is affected mainly by collisions between the metal atoms themselves, rather than by diffusion through an aggregate of non-diffusing gas molecules. Therefore mass transfer occurs by bulk flow of metal atoms between the evaporator and condenser, with entrainment of any minor evaporating species in the principal flow. The effective back pressure immediately above the evaporator will be somewhat less than that of the same molecular concentration exerting a pressure equal in all directions. It is suggested that if the average net velocity of zinc vapour across the distillation space be v cm/sec and the average projected velocity of zinc atoms be $\bar{c}$ cm/sec then, since $p \propto (\bar{c})^2$, the effective pressure changes from p_1 in the absence of flow to $(1-r)^2 p_1$, where $r = v/(\bar{c})$. The corresponding expression for the effective pressure of zinc adjacent to the condenser is $(1+r)^2 p_2$, where p_2 would be the pressure in the absence of flow.

Based on the concepts mentioned above, combined with the Langmuir expressions for vacuum distillation, the net rates of evaporation E and condensation C can be written in terms of the area A_e and A_c and accommodation coefficients α_e and α_c of the evaporating and condensing surfaces as,

$$E = \alpha_e A_e a \{p^* - p_1(1-r_1)^2\} \qquad (5\text{-}74)$$

$$C = \alpha_c A_c \{a(1+r_2)^2 p_2 - b p_c\} \qquad (5\text{-}75)$$

where p^* and p_c refer to the equilibrium vapour pressures, p_1 and p_2, the pressure immediately adjacent to the surfaces and a and b are the Langmuir terms $(M/2\pi RT)^{1/2}$ and $(M/2\pi RT_c)^{1/2}$ for the temperatures T and T_c at

蒸留ギャップ間（蒸発面から凝集面までの距離）における蒸留金属原子の移動は、拡散に関与しないガス分子間の拡散より、蒸留される金属分子間の衝突により影響を受ける。このようにして蒸発器と凝集器間の金属原子の物質移動は進行し、他に微量の蒸発元素が存在する場合、この元素はこの流れに取り込まれることになる。

蒸発面直上の圧力は、全方向に同じ濃度の蒸発分子を持つ（流れの影響が無い）場合の圧力より小さくなると考えられる。蒸留ギャップ間の亜鉛の移動速度がv、平均速度が$\bar{c}$である場合、流れがない場合の圧力p_1に対して、流れの影響を受ける場合の圧力は$(1-r)^2 p_1$となる。ここでr = v/$\bar{c}$である。

同様に、流れがある場合の凝集面直上の圧力は$(1+r)^2 p_2$となる。ここでp_2は流れがない場合の圧力である。

以上の考えに基づき、蒸発による移動量Eと凝集による移動量Cは、それぞれの有効面積（A_e、A_c）や適応係数（α_e、α_c）を用いて左記の(5-74)、(5-75)式のように表すことができる。

p^*、p_cはそれぞれ蒸発面直上、凝集面直上の平衡蒸気圧であり、p_1、p_2はそれぞれ蒸発面、凝集面直上の平衡蒸気圧である。

また、a、bはそれぞれのラングミュア項、T、T_cは蒸発器と凝集器の温度である。

蒸留器内のガス流れの平均値には変動があるが、凝集面直近に到るまでは明確な温度の変化は無いとみなされる。また、真空脱亜鉛における条件下では、圧力ヘッドに比べて流速ヘッドの変化は顕著でないと考えられる。このように、摩擦によるエネルギー散逸がないと仮定すると、p_1とp_2は等しいとみすことができる。

液相中の亜鉛の物質移動係数、バルクの亜鉛濃度、界面の亜鉛濃度を用いて、蒸留される亜鉛の表面（真空／液相界面）への移動速度（モル流束）は(5-76)式のように表される。

evaporator and condenser respectively.

Although changes in average vapour velocity occur in the flow from evaporator to condenser, an appreciable change in gas temperature would not be expected until the immediate vicinity of the condensing surface is reached. Also, from an energy balance on a stream tube for the conditions likely to be met in vacuum dezincing, changes in velocity head are insignificant compared with the pressure head; thus, provided there is no dissipation of energy by friction, the pressures p_1 and p_2 would be essentially equal.

In terms of a liquid-phase mass transfer coefficient k_{Zn} and concentrations C_{Zn}^b and C_{Zn}^* in the bulk and surface respectively, the rate of supply of the distilling species is given by

$$\frac{\dot{n}_{Zn}}{A_e} = k_{Zn}(C_{Zn}^b - C_{Zn}^*) \qquad (5\text{-}76)$$

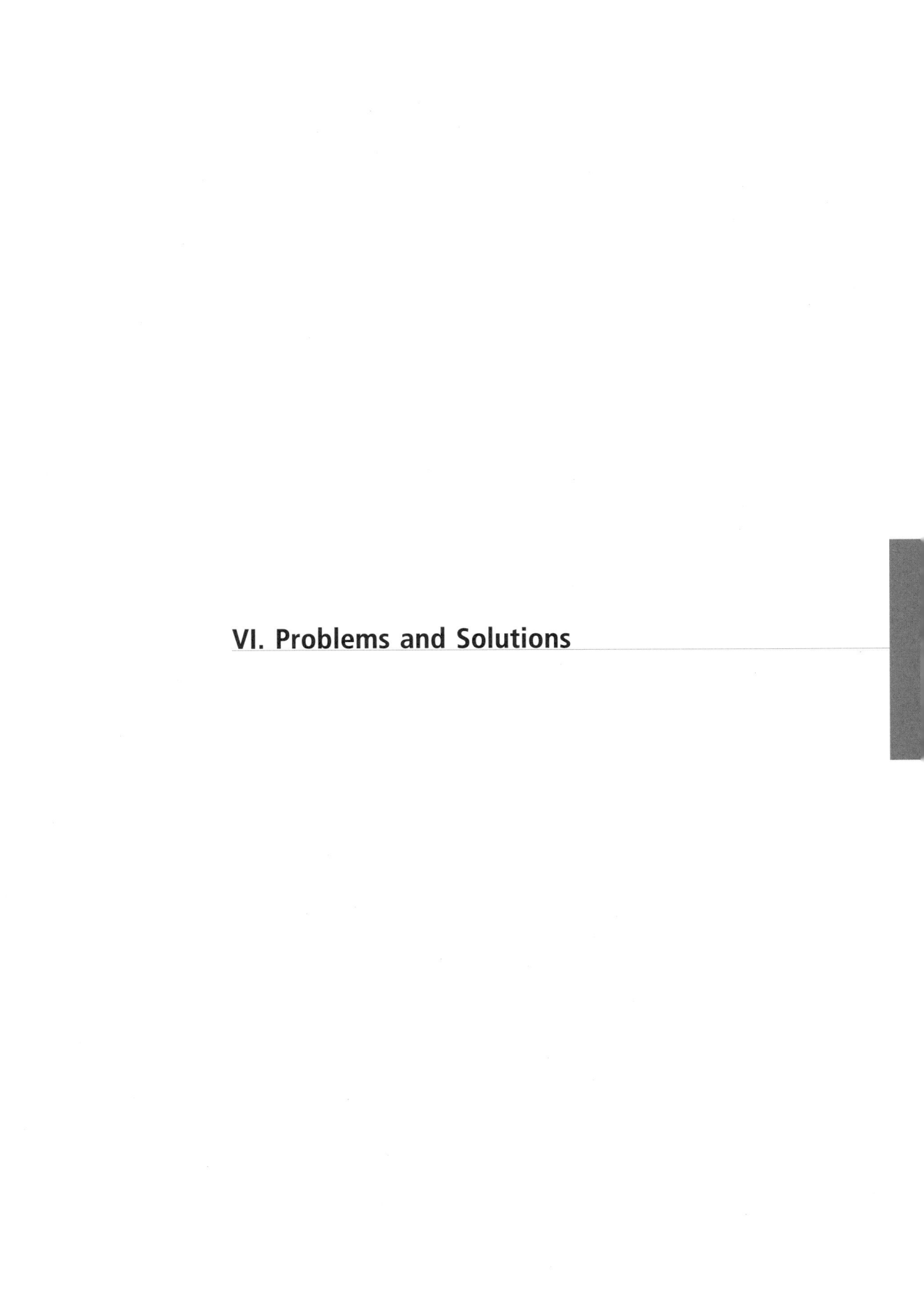

VI. Problems and Solutions

Problem 1
The Diffusion Coefficient of Zinc Vapor in Nitrogen

In the Imperial Smelting process zinc is reduced from its oxide in a lead-zinc blast furnace by CO gas, and is carried into the upper regions of the furnace as a vapor. The top gas composition typically is

  Zn   5%
  CO_2 7 to 10%
  CO   25%
  N_2 balance

The zinc vapor is then shock quenched in a lead splash condenser.

An apparatus has been constructed to measure the diffusion coefficient of zinc in nitrogen in order that some diffusion calculations can be performed on the splash condenser. The apparatus is shown in Fig.6-1-1.

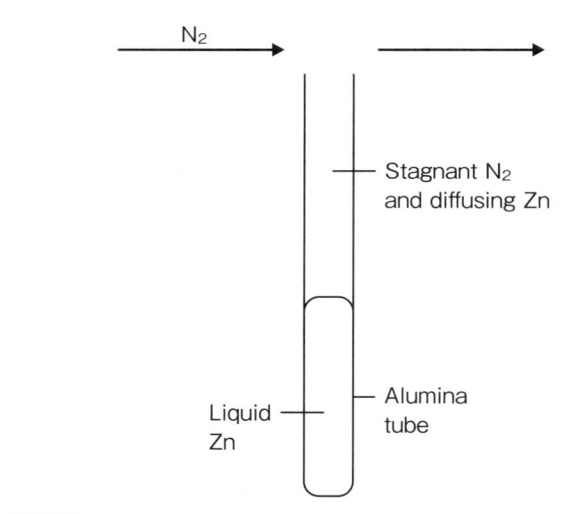

Fig. 6-1-1 Experimental apparatus for measurement of the diffusion coefficient of Zn vaper in nitrogen.

A 2mm i.d. alumina tube containing liquid zinc is positioned inside a furnace which is held at a constant temperature of 800℃. The distance from the surface of the liquid zinc to the top of the tube is 4.0×10^{-2} m. Nitrogen gas is passed through the furnace and across the top of the tube at a high flow rate. In the course of an experiment, it is found that the level of zinc in the tube drops by 2.0×10^{-3} m in 24 hours. If the nitrogen in the space above the zinc is stagnant, what is the value of the interdiffusion coefficient, D_{Zn,N_2}?

Pressure inside the tube is 1.013×10^5 Pa.

R, the gas constant, is $8.314 \, Pa \cdot m^3 \cdot K^{-1} \cdot mole^{-1}$

The equilibrium vapor pressure of zinc, p_{Zn} (Pa), is expressed as a function of temperature, T(K).

$$\ln p_{Zn} = \frac{-14.183 \times 10^3}{T} + 23.57 \quad (6\text{-}1\text{-}1)$$

The density of zinc at 800℃ is 6.40×10^{-3} kg m^{-3}.

The molecular weight of zinc is 6.50×10^{-2} kg mole^{-1}.

鉛表面の管上部からの距離は4cmである。管上部は開口されており、上部には十分な流量の窒素が流れている。

この状態で24時間保持したところ液体亜鉛表面が0.2cm低下した。

アルミナ管中の亜鉛蒸気を含む窒素ガスは淀んでいるものとして、窒素中の亜鉛ガスの拡散係数を求めよ。

実験条件や関係諸量は左記のように与えられている。

なお、亜鉛蒸気圧の温度依存性は(6-1-1)式で与えられる。

Problem 1 The Diffusion Coefficient of Zinc Vapor in Nitrogen

Problem 1 Solution

図6-1-1の装置を図6-1-2のように模式化する。

アルミナ管内部に、外部の気体の流れによる対流は生じないものとする。

亜鉛ガスのモル流束は(6-1-2)式に示すように、液体亜鉛の表面レベル変化と関連付けられる。

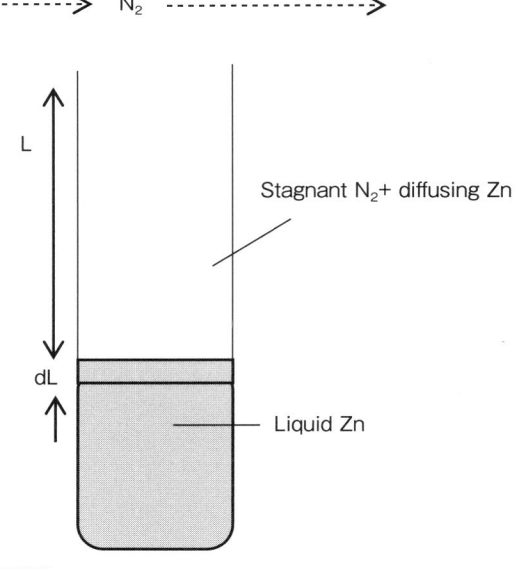

Fig. 6-1-2 Model description for mesurement of diffusion coefficient of zinc in nitrogen.
L : Exposed length of tube above liquid level.
dL : Drop in liquid level in time increment, dt.
Assumptions : No convective mixing in the tube in the space above the liquid zinc.

The molar flux $\dfrac{\dot{n}_{Zn}}{A}$ can be related to the change in liquid level dL, through the following relation.

$$\frac{\dot{n}_{Zn}}{A} = C_{Zn(l)} \frac{dL}{dt} \qquad (6\text{-}1\text{-}2)$$

where $C_{Zn(l)}$ is a molar density of liquid zinc.

<Note>

For small changes (dL) it appears that the mass flux can also be calculated by considering the total loss in weight over the time period. The answer does not appear to be very different from the one that will be obtained by the above approach, although this solution is mathematically more correct.

dLの値が非常に小さいためマスフラックスは重量の全変化と関連付ける事が出来き、その結果は本解答と大差はないが、本解答の方が理論的にはより正しいアプローチと言える。

Recall the relation which was derived for the case of diffusion of component⟨1⟩ through stagnant⟨2⟩. This equation (6-1-3) was derived for steady state which may be assumed for this 24 hr experiment.

$$\frac{\dot{n}_1}{A} = D_{1,2} C_T \frac{(X_{1,o} - X_{1,L})}{L X_{2,\ln}} \quad (6\text{-}1\text{-}3)$$

Equating the two expressions obtained for molar flux,

$$C_{Zn(l)} \frac{dL}{dt} = D_{Zn,N_2} C_T \frac{X_{Zn,o} - X_{Zn,L}}{L X_{N_2,\ln}} \quad (6\text{-}1\text{-}4)$$

$X_{Zn,o}$ --- mole fraction of Zn at the liquid surface

$X_{Zn,L}$ --- mole fraction of Zn at the top of the alumina tube

C_T --- molar density of gas phase

$$\int_{L_o}^{L_f} L\,dL = \frac{D_{Zn,N_2} C_T (X_{Zn,o} - X_{Zn,L})}{C_{Zn(l)} X_{N_2,\ln}} \int_o^t dt \quad (6\text{-}1\text{-}5)$$

$$\frac{L_f^2 - L_o^2}{2} = \frac{D_{Zn,N_2} C_T (X_{Zn,o} - X_{Zn,L}) t}{C_{Zn(l)} X_{N_2,\ln}} \quad (6\text{-}1\text{-}6)$$

$$\therefore D_{Zn,N_2} = \frac{C_{Zn(l)} X_{N_2,\ln} (L_f^2 - L_o^2)}{2 C_T (X_{Zn,o} - X_{Zn,L}) t} \quad (6\text{-}1\text{-}7)$$

$$C_{Zn(l)} = \frac{6.4 \times 10^3}{6.5 \times 10^{-2}} \frac{\text{kg·m}^{-3}}{\text{kg·mole}^{-1}} = 9.85(10^4) \text{ mole/m}^3$$

$$C_T = \frac{n}{V} = \frac{P}{RT} \text{ (assuming ideal gas)}$$

$$= \frac{101300 \text{Pa}}{8.314 \frac{\text{m}^3 \cdot \text{Pa}}{\text{mole·K}} 1073 \text{K}} = 11.3 \text{mole/m}^3$$

Partial pressure of Zn calculated from the eq.(6-1-1) in the problem is 31194 Pa.

$$X_{Zn,o} \text{ (at surface)} = \frac{31194}{101300} = 0.308$$

$X_{Zn,L}$ (at top) = 0 (N$_2$ sweeps Zn away)

$X_{N_2,L} = 1.0$, $X_{N_2,o} = 0.692$

$$\therefore X_{N_2,\ln} = \frac{0.308}{\ln \frac{1.0}{0.692}} = 0.837$$

$t = 24$hr or $8.64(10^4)$sec

一方、本実験の系のように淀んでいる長さLの成分⟨2⟩の層中を拡散する成分⟨1⟩のモル流束（$\frac{\dot{n}_1}{A}$）は(6-1-3)式で与えられる。

（注）(6-1-3)式の導出については、2章2-1節、あるいは"Transport Phenomena Second Edition"[1]の§18.2を参照のこと。

(6-1-2)式と(6-1-3)式より(6-1-4)式が得られる。

(6-1-5)式をLについてL$_0$からL$_f$まで、tについて0からtまで積分して(6-1-6)式を得る。

これに諸数値を代入してD$_{Zn,N_2}$の値を0.224cm^2/secと求めることができる。

$$D_{Zn,N_2} = \frac{9.85 \times 10^4 \times 0.837(4.2^2 - 4.0^2) \times 10^{-4}}{2 \times 11.3 \times 0.308 \times 8.64 \times 10^4}$$

$$= 0.2248 \times 10^{-4} \text{ m}^2/\text{sec}$$

Problem 2
The Roasting of Zinc Sulphide with Oxygen

Roasting is usually an important first step in the extraction of metals from sulphide ores. In the case of zinc processing, particles of zinc sulphide are roasted by air to form zinc oxide, that is,

$$ZnS(s) + 3/2\, O_2(g) = ZnO(s) + SO_2(g) \quad (6\text{-}2\text{-}1)$$

It is known, from thermodynamic calculations, that this reaction proceeds completely to products when at equilibrium. Kinetic studies further have shown that a given particle of zinc sulphide is roasted in a topochemical fashion.

(1) Describe what this (a topochemical fashion) means and, in words, elaborate on the steps which may govern the rate of the roasting reaction.

In one kinetic investigation, a small (2cm × 2cm) plate of zinc sulphide, enclosed inside a furnace, was suspended from the arm of a balance (vide Fig.6-2-1).

「焙焼」は硫化鉱石から金属を抽出する第1段階である。亜鉛製錬の場合、硫化亜鉛の粒子は (6-2-1) 式に示すように空気によって焙焼されて酸化亜鉛となる。

熱力学的な考察によって、(6-2-1) 式の反応は瞬時に進行する事が知られている。また、速度論的な研究によって粒子状の硫化亜鉛の粒の焙焼は"トポケミカル"状に進行するすることも知られている。

(1) この"トポケミカル"状に反応が進行する"とはどのような意味なのか説明せよ。

図6-2-1に示すように、2cm角の硫化亜鉛の板を加熱炉中に設置し、熱天秤によって重量変化を測定する。
炉の中は酸素ガスが一定の流量流れており、焙焼反応が進むにつれて変化するサンプル重量の減少より反応速度が測定される。
850℃での実験において反応速度 ($\dot{n}_{ZnS}$) は 7.18×10^{-3} (モル/分) であった。

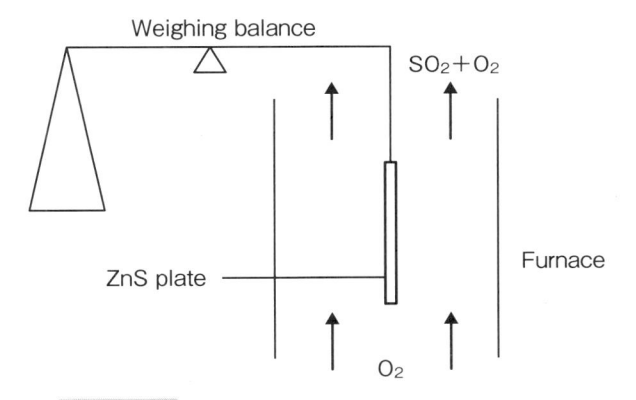

Fig. 6-2-1　Schematic of experimental apparatus.

Oxygen was passed through the furnace, and as the reaction proceeded, the weight of the sample was measured continuously. From the weight loss curve the reaction rate was determined. During one experiment at 850°C, the rate of reaction, $\dot{n}_{ZnS}$, was found to be $7.18(10^{-3})$ g-mole per minute just after the oxygen was first introduced into the furnace.

(2) If at this early reaction time, it can be assumed that the ZnO layer is negligibly thin (it offers no resistance to the reaction rate), show that the rate of roasting is controlled by diffusion in the boundary layer in the gas phase. Assume that the edges of the ZnS plate have a negligible effect on the reaction rate.

Pressure of oxygen in the bulk of the gas stream = 1.0 atm.
Diffusion coefficient for the O_2/SO_2 binary gas mixture at 850°C = 1.73 cm² sec⁻¹.
Gas constant = 82.1 cm³ atm mole⁻¹ K⁻¹.
Average thickness of the gas boundary layer = 1.0 cm.
Molecular weight of ZnS = 97.5 g mole⁻¹.

(3) Later on in the experiment, would the rate of reaction increase or decrease with time?

Problem 2 Solution

(1) "Topochemical reaction" means a chemical reaction which takes place at the boundary of solid phases.

At the first stage of the reaction, the step which controls the rate of reaction may be diffusion in the boundary layer between solid and gas.

However, once the product layer is formed, the controlling step changes to the diffusion in the product layer (ZnO) at the surface.

If we cut the solid particle at the cirtain period, the concentrically circular product layer will be found around the core which have not reacted yet.

(2) If we assume ① the fast heterogeneous reaction and ② reaction control lies in the transport of gaseous reactant to the interface, reaction rate is expressed as follows.

$$gX + G_1 = fG_2 + hY \qquad (6\text{-}2\text{-}2)$$

$$\frac{\dot{n}_{G_1}}{A} = \frac{D_{G_1,G_2}C_T}{(1-f)\delta} \ln \frac{1-(1-f)X_{G_1,\delta}}{1-(1-f)X_{G_1,0}} \qquad (6\text{-}2\text{-}3)$$

Regarding the reaction of zinc sulphide with oxygen,

$$ZnS(s) + \frac{3}{2}O_2(g) = ZnO(s) + SO_2(g) \qquad (6\text{-}2\text{-}4)$$

$f = \frac{2}{3}$ (—)

$A = 4 \times 2$ (cm^2)

$D_{O_2} = 1.73$ ($\frac{cm^2}{sec}$)

$C_T = \frac{P^0_{O_2}}{R \cdot T}$ ($\frac{mole}{cm^3}$)

$P^0_{O_2} = 1$ (atm)

$R = 82.1$ (cm$^3 \cdot \frac{atm}{mole \cdot K}$)

(1) トポケミカル反応において、初期段階では反応の律速段階は固相と気相の境界層であるが、一旦反応層が形成されると律速段階はこの表面に形成された反応層となる。一定の時間が経過した後、固体粒子を切断すると、表面は周に沿った反応層が形成され、内部は未反応である断面を観察することが出来る。

(2) 第2章の第2節で説明したように、非等モル拡散((6-2-2)式)におけるモル流束は(6-2-3)式となる。ZnSの焙焼反応の式(6-2-4)と(6-2-2)式を対比させると各係数が求まり、与えられた条件を入れて$\dot{n}_{O_2}$および$\dot{n}_{ZnS}$が求まる。

$$\delta = 1.0 \text{ (cm)}$$

From the assumptions

$$X_{O_2,\delta} = \frac{P_{O_2,\delta}}{P_{O_2}^0} = 1.0 \, (-)$$

$$X_{O_2,0} = \frac{P_{O_2,0}}{P_{O_2}^0} = 0 \, (-)$$

Therefore

$$\dot{n}_{O_2} = \frac{8 \times 1.73 \times \left(\frac{1}{82.1} \times 1123\right)}{\left(1 - \frac{2}{3}\right) \times 1} \ln \frac{1 - \left(1 - \frac{2}{3}\right) \times 1}{1 - \left(1 - \frac{2}{3}\right) \times 0}$$

$$= -1.826 \times 10^{-4} \, \left(\frac{\text{mole}}{\text{sec}}\right)$$

$$\dot{n}_{ZnS} = \frac{2}{3} |\dot{n}_{O_2}| = 1.217 \times 10^{-4} \, \left(\frac{\text{mole}}{\text{sec}}\right)$$

Since $\dot{n}_{ZnS}$ was found to be $7.18 \times 10^{-3} \left(\frac{\text{mole}}{\text{min}}\right)$, which is $1.20 \times 10^{-4} \left(\frac{\text{mole}}{\text{sec}}\right)$, the diffusion of O_2 in the boundary layer can be the rate controlling step.

実際に測定された硫化亜鉛の反応速度は 7.18×10^{-3} (mole/min)（$=1.20 \times 10^{-4}$ (mole/sec)）であり、計算値$\dot{n}_{ZnS}$とよく一致して、モデルの前提条件としたとおり境界層における酸素の拡散が律速段階であることがわかる。

一旦、硫化亜鉛粒子表面に反応層が形成されると、これが律速段階となる。ここでの拡散係数は気相の拡散係数に比べ小さい事は明らかであり、反応速度は次第に低下すると予測される。

Problem 3
Fume Formation in basic oxgen furnace

Brown fumes of iron oxide are formed whenever iron is converted into steel; and, in the pneumatic steelmaking processes, such as the BOF, the quantity of such fume is especially large. The fumes occur when iron vapor, close to the liquid steel surface, reacts with oxygen in the gas phase. Fumes are an undesirable feature of these processes for a number of reasons: serious pollution, loss of metal and refractory attack in the furnace.

F company has launched an investigation into the causes of fume formation as a first step toward fume control in its BOF operation. Their research engineers have performed a series of isothermal experiments at 1600℃ in which an oxidizing gas of argon and oxygen is passed over a pool of liquid iron. It is found that the rate of iron loss as fume depends on the first power of the partial pressure of oxygen up to a critical partial pressure, after which no iron as fume is lost, vide Fig. 6-3-1.

The engineers are unable to explain their results so

第1章で示したBOF（転炉）プロセスなど、酸素を用い銑鉄を鋼に精錬するプロセスにおいては酸化鉄のフュームが発生する場合がある。このフュームは溶鋼の表面近くで鉄蒸気が気相の酸素と反応することによって生じる。フュームの発生は、大気汚染、メタルのロス、および炉内耐火物の損傷など数々の理由で好ましいものではない。

F社は転炉の操業を始めるに際しフューム発生制御の第一ステップとしての研究に着手した。F社の研究者達は1600℃の溶けた銑鉄のプール上にAr＋酸素ガスを流す基礎実験を行い、フュームとしての鉄ロスの速度は、ある酸素分圧まではこの酸素分圧の1次に比例するが、それ以上の分圧ではロスは0となる結果を得た（図6-3-1）。

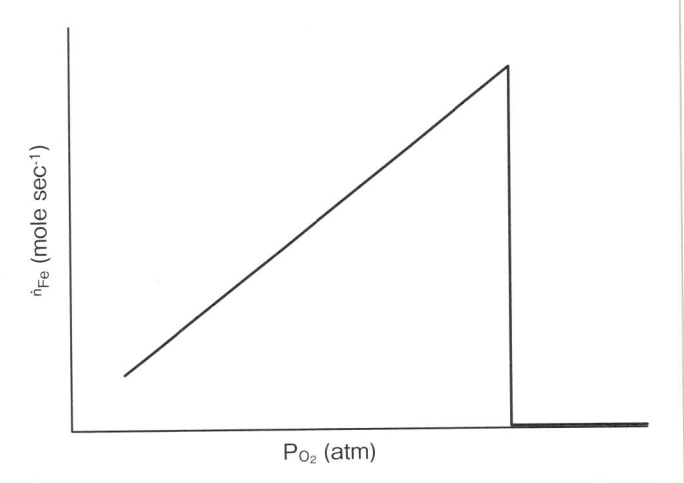

Fig. 6-3-1 The effect of P_{O_2} on the rate of iron loss, $\dot{n}_{Fe}$.

しかしながら彼らはこの実験結果を説明することが出来なかった。物質移動の基礎および鉄の蒸発速度の式((6-3-1)式)から鉄ロスの速度と酸素分圧の関係を解明せよ。

加えて、フューム発生の最大速度を数学的に求めると共に、今後の精錬操業に向けてフュームの発生を抑制する方法を提案せよ。

you have been called in to consult. Using mass transfer concepts and the following equation,

$$\frac{\dot{n}_{Fe}}{A} = \frac{k_{ev}}{\sqrt{RT}}(P^*_{Fe} - P^0_{Fe}) \qquad (6\text{-}3\text{-}1)$$

where $\dfrac{\dot{n}_{Fe}}{A}$ is the flux of iron evaporating from the surface

k_{ev} is an evaporation constant

P^*_{Fe} is the equilibrium vapor pressure of iron

P^0_{Fe} is the partial pressure of iron at the surface,

justify the relation between rate of iron loss and P_{O_2}. Part of your job also is to determine mathematically the maximum rate of fume formation and finally to suggest ways of reducing fume in future steelmaking operations.

Problem 3 Solution

A number of steps are involved in the formation of brown fume.

(1) Iron evaporates from the surface of the liquid metal. The flux of iron evaporated is

$$\left(\frac{\dot{n}_{Fe}}{A}\right)_{ev} = \frac{K_{ev}}{\sqrt{RT}}(P_{Fe}^* - P_{Fe}^0) \qquad (6\text{-}3\text{-}1)$$

(2) Iron vapor diffuses away from the liquid metal surface where the partial pressure of iron is greatest. The diffusive flux* of iron(v) is

$$\frac{\dot{n}_{Fe}}{A} = -D_{Fe,g} C_T \frac{dX_{Fe}}{dy} \qquad (6\text{-}3\text{-}2)$$

or

$$\frac{\dot{n}_{Fe}}{A} = -\frac{D_{Fe,g}}{RT} \frac{dP_{Fe}}{dy} \qquad (6\text{-}3\text{-}3)$$

since

$$C_T X_{Fe} = C_{Fe} \qquad (6\text{-}3\text{-}4)$$

and

$$C_{Fe} = \frac{P_{Fe}}{RT} \text{ (assuming gas is ideal)} \qquad (6\text{-}3\text{-}5)$$

(3) Oxygen diffuses from the bulk of the gas stream toward the surface of the liquid iron. The molar flux* of oxygen is

$$\frac{\dot{n}_{O_2}}{A} = \frac{-D_{O_2,g}}{RT} \frac{dP_{O_2}}{dy} \qquad (6\text{-}3\text{-}6)$$

*We will assume that the concentration of O_2 and Fe vapor are small enough that bulk movement of the gas due to diffusion is negligible.

(4) The oxygen and iron vapor react in the gas phase to yield FeO.

$$Fe + \frac{1}{2}O_2 = FeO \qquad (6\text{-}3\text{-}7)$$

These steps are shown schematically in Fig. 6-3-2.

フュームの生成反応には多くのステップが含まれている。

(1) 溶鉄表面からの鉄の蒸発速度は(6-3-1)式で与えられている。

(2) 鉄蒸気は最大の分圧となる溶鉄表面から気相に向けて拡散するが、そのモル流束は(6-3-2)式、または(6-3-3)式で表される。

(3) 酸素は気体バルクから溶銑プール表面に向けて拡散するが、そのモル流束は(6-3-6)式で表される。

ここで、酸素や鉄蒸気の濃度は十分に低いため、拡散に基づく気体バルクの流れは無視できるものとする。

(4) 酸素や鉄蒸気は気相中で反応してFeOを生成する。

これらのステップは図6-3-2に図示されている。

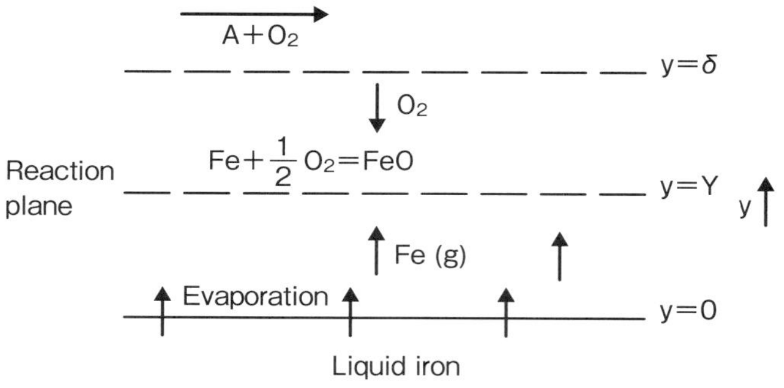

```
                    A+O₂ →
  ─── ─── ─── ─── ─── ─── ─── ─── y=δ
                      ↓ O₂
Reaction    Fe + ½ O₂ = FeO
plane   ─── ─── ─── ─── ─── ─── ─── y=Y    y↑
                      ↑ Fe (g)
      ↑ Evaporation ↑         ↑
  ─── ─── ─── ─── ─── ─── ─── ─── y=0
                Liquid iron
```

Fig. 6-3-2 Schematic of iron oxide fume formation.

これらの式で示した現象は厚みδの境界層中で起こると考える。

第2章 第3節で均一反応を伴う拡散、特に亜鉛蒸気の酸化について述べたが、化学反応速度が速く、不可逆の場合、反応を記述する数式は単純化されることを示した。

反応が速いということは、反応面では平衡に近い状態が達成されていると考えることができ、その面における反応種と生成種の濃度の推定が熱力学を通して可能なことになる。さらに不可逆反応であれば反応物は反応面では存在せず、その濃度は0とみなすことができる。このことは、熱力学的に見ると反応が大きな負の自由エネルギーを持ち、平衡定数((6-3-8)式)は非常に大きいことを意味する。(6-3-8)式からも平衡定数が非常に大きい場合、反応物の活量（濃度）は0に近いことがわかる。

反応の速度や不可逆性は同時に均一反

We are considering that the above events all occur within a boundary layer of thickness δ.

We have, in class, examined the diffusion of reacting species which are involved in an homogeneous chemical reaction; in particular we discussed the oxidation of zinc vapor. We found that the mathematics are simplified considerably if the chemical reaction is very fast and practically "irreversible" (goes all the way to products). The fact that the reaction is fast means that equilibrium is closely approached, and thus we can predict the equilibrium concentrations of reactants and products in the reaction zone, from the thermodynamics.

The "irreversibility" of the reaction readily suggests that the concentrations of the reactants in the reaction zone are nearly zero. Thus the reactants cannot coexist. That this is true can be found from the thermodynamics since the reaction we are describing is one with a large negative Gibbs free energy, or a large equilibrium constant. The equilibrium constant is defined as

$$K = \frac{a_m \cdot a_n \text{ (Products)}}{a_i \cdot a_j \text{ (Reactants)}} \quad (6\text{-}3\text{-}8)$$

which if very large means very low activities (or concentrations, if activity coefficients are normal) of the reactants.

The speed and irreversibility of a reaction also means

that the homogeneous reaction will occur at a definite plane, the position of which depends on variables like bulk concentrations of reactants.

The reaction between Fe and O_2 is probably fast and is certainly quite irreversible. It is most likely fast because it is occurring at 1600 ℃, a temperature at which the activation energy for reaction would have to be very large.

For these reasons we draw a reaction plane at y=Y in Fig. 6-3-2. It is clear that the concentrations of O_2 and Fe at the plane are very small; also it is apparent that no O_2 is present in the gas below the plane $0<y<Y$, and no Fe vapour exists above it, $Y<y<\delta$.

In order to solve the first part of the problem, we will combine eqs. (6-3-1), (6-3-2), and (6-3-6), to give an overall rate equation.

Consider first eq. (6-3-6) for the diffusion of O_2 above the FeO reaction plane. Since at steady state $\frac{d}{dy}\left(\frac{\dot{n}_{O_2}}{A}\right)=0$, the equation can be integrated directly between the limits:

$$(1)\quad y=\delta,\ P_{O_2}=P_{O_2,\delta} \qquad (6\text{-}3\text{-}9)$$

$$(2)\quad y=Y,\ P_{O_2}=P_{O_2,Y} \qquad (6\text{-}3\text{-}10)$$

(Note that $P_{O_2,\delta}$ is the bulk oxygen partial pressure, while $P_{O_2,Y}=0$ because K for the FeO reaction is very large, and equilibrium is assumed).

$$\therefore \frac{\dot{n}_{O_2}}{A}\int_\delta^Y dy = \frac{-D_{O_2,g}}{RT}\int_{P_{O_2,\delta}}^0 dP_{O_2} \qquad (6\text{-}3\text{-}11)$$

$$\therefore \frac{\dot{n}_{O_2}}{A} = \frac{-D_{O_2,g}}{(\delta-Y)}\frac{P_{O_2,\delta}}{RT} \qquad (6\text{-}3\text{-}12)$$

Similarly for the case of iron vapour diffusion below the reaction plane, eq. (6-3-2) can be integrated between the limits.

$$y=0,\ P_{Fe}=P_{Fe,0} \qquad (6\text{-}3\text{-}13)$$

$$y=Y,\ P_{Fe}=P_{Fe,Y}=0 \qquad (6\text{-}3\text{-}14)$$

$$\therefore \frac{\dot{n}_{Fe}}{A}\int_0^Y dy = \frac{-D_{Fe,g}}{RT}\int_{P_{Fe,0}}^0 dP_{Fe} \qquad (6\text{-}3\text{-}15)$$

応が限られた場所で起こることを意味し、その位置はバルク中の濃度などの変数に依存している。

鉄と酸素の反応も非常に速く、不可逆である。なぜならばこの反応は1600℃という高温で進行し、活性化エネルギーは非常に大きいからである。

以上述べたように図6-3-2においてy=Yでの反応面を仮定する。反応面における鉄や酸素の濃度は非常に低いと予測できる。また反応面以下では酸素は存在せず同様に、その面より上では鉄は存在しないとみなして良い。
(6-3-1)、(6-3-2)、および(6-3-6)式を組み合わせて、総括速度式を求める。まずはじめに反応面より上部の酸素の拡散について考察する。定常状態においては $\frac{d}{dy}\left(\frac{\dot{n}_{O_2}}{A}\right)=0$ であり、この式を(6-3-9)、(6-3-10)式の境界条件の下に積分すると (6-3-12) 式が得られる。

同様に反応面より下部の鉄蒸気の拡散については (6-3-2) 式を (6-3-13)、(6-3-14)式の境界条件の下に積分して(6-3-16)式を得る。

次に、(6-3-1)式と(6-3-16)式とを組み合わせ変形して(6-3-21)式を得る。

$$\therefore \frac{\dot{n}_{Fe}}{A} = \frac{D_{Fe,g}}{Y} \frac{P_{Fe,0}}{RT} \quad (6\text{-}3\text{-}16)$$

Now we combine eq. (6-3-1) and eq. (6-3-16) by first rearranging as follows

$$\left(\frac{\dot{n}_{Fe}}{A}\right)_{ev} \frac{\sqrt{RT}}{K_{ev}} = P_{Fe}^* - P_{Fe,0} \quad (6\text{-}3\text{-}17)$$

$$\frac{\dot{n}_{Fe}}{A} \frac{Y\,RT}{D_{Fe,g}} = P_{Fe,0} \quad (6\text{-}3\text{-}18)$$

Then adding and utilizing the fact that

$$\left(\frac{\dot{n}_{Fe}}{A}\right)_{ev} = \frac{\dot{n}_{Fe}}{A} \quad (6\text{-}3\text{-}19)$$

we obtain

$$\left(\frac{\dot{n}_{Fe}}{A}\right)_{ev} \left\{\frac{\sqrt{RT}}{K_{ev}} + \frac{YRT}{D_{Fe,g}}\right\} = P_{Fe}^* \quad (6\text{-}3\text{-}20)$$

and

$$\left(\frac{\dot{n}_{Fe}}{A}\right)_{ev} = \left\{\frac{1}{\frac{\sqrt{RT}}{K_{ev}} + \frac{YRT}{D_{Fe,g}}}\right\} P_{Fe}^* \quad (6\text{-}3\text{-}21)$$

(6-3-21)式中のYは変数であり、一般的な総括速度式を得る上で、できれば消去したいものである。そのために(6-3-22)式で示される化学量論の関係を用いる。

Y in eq. (6-3-21) is a variable which preferably should be eliminated to obtain a general, overall rate expression. To accomplish this we make use of the stoichiometry of the reaction

$$\left(\frac{\dot{n}_{Fe}}{A}\right)_{ev} = \frac{-2\dot{n}_{O_2}}{A} \quad (6\text{-}3\text{-}22)$$

(6-3-12)式と(6-3-21)式より(6-3-23)式を得る。

and eqs. (6-3-12) and (6-3-21) accordingly.

$$\left\{\frac{1}{\frac{\sqrt{RT}}{K_{ev}} + \frac{YRT}{D_{Fe,g}}}\right\} P_{Fe}^* = \frac{2D_{O_2,g}}{\delta - Y} \frac{P_{O_2,\delta}}{RT} \quad (6\text{-}3\text{-}23)$$

$$\frac{K_{ev} D_{Fe,g} P_{Fe}^*}{D_{Fe,g} \sqrt{RT} + Y K_{ev} RT} = \frac{2D_{O_2,g} P_{O_2,\delta}}{(\delta - Y) RT} \quad (6\text{-}3\text{-}24)$$

$$(\delta - Y) RT\, K_{ev} D_{Fe,g} P_{Fe}^* = 2(D_{Fe,g}\sqrt{RT} + Y K_{ev} RT) D_{O_2,g} P_{O_2,\delta} \quad (6\text{-}3\text{-}25)$$

$$2YK_{ev}RTD_{O_2,g}P_{O_{2\delta}} + YRT\,K_{ev}D_{Fe,g}P_{Fe}{}^* = \delta RT\,K_{ev}D_{Fe,g}P_{Fe}{}^* - 2D_{Fe,g}\sqrt{RT}\,D_{O_2,g}\,P_{O_{2,\delta}} \tag{6-3-26}$$

$$\therefore Y = \frac{\delta RT\,K_{ev}D_{Fe,g}P_{Fe}{}^* - 2D_{Fe,g}\sqrt{RT}\,D_{O_2,g}\,P_{O_{2,\delta}}}{2K_{ev}RT\,D_{O_2,g}P_{O_{2,\delta}} + RT\,K_{ev}D_{Fe,g}P_{Fe}{}^*} \tag{6-3-27}$$

$$Y = \delta - \frac{\dfrac{2D_{O_2,g}\,P_{O_{2,\delta}}}{K_{ev}\sqrt{RT}\,P_{Fe}}}{1 + \dfrac{2D_{O_2,g}P_{O_{2,\delta}}}{D_{Fe,g}P_{Fe}{}^*}} \tag{6-3-28}$$

Substitute eq. (6-3-27) into eq. (6-3-21)、

$$\left(\frac{\dot{n}_{Fe}}{A}\right)_{ev} = \left\{\frac{1}{\dfrac{\sqrt{RT}}{K_{ev}} + \dfrac{\delta RT\,K_{ev}P_{Fe}{}^* + 2\sqrt{RT}\,D_{O_2,g}P_{O_{2,\delta}}}{2K_{ev}D_{O_2,g}P_{O_{2,\delta}} + K_{ev}D_{Fe,g}P_{Fe}{}^*}}\right\}P_{Fe}{}^* \tag{6-3-29}$$

(6-3-27)式を(6-3-21)式に代入して変形すると(6-3-31)式を得る。

$$\left(\frac{\dot{n}_{Fe}}{A}\right)_{ev} = K_{ev}P_{Fe}{}^*\left\{\frac{2D_{O_2,g}P_{O_{2,\delta}} + D_{Fe,g}P_{Fe}{}^*}{2\sqrt{RT}\,D_{O_2,g}P_{O_{2,\delta}} + \sqrt{RT}\,D_{Fe,g}P_{Fe}{}^* + \delta RT\,K_{ev}P_{Fe}{}^*} - 2\sqrt{RT}\,D_{O_2,g}P_{O_{2,\delta}}\right\} \tag{6-3-30}$$

$$\left(\frac{\dot{n}_{Fe}}{A}\right)_{ev} = K_{ev}\left(\frac{2D_{O_2,g}P_{O_{2,\delta}} + D_{Fe,g}P_{Fe}{}^*}{\sqrt{RT}\,D_{Fe,g} + \delta RT\,K_{ev}}\right) \tag{6-3-31}$$

Thus we can see from eq. (6-3-31) that the rate of evaporation is a linear function of $P_{O_{2,\delta}}$ as shown in the graph.

To understand why there is an abrupt decrease in the evaporation rate we turn to eq. (6-3-28). Here we see that $P_{O_{2,\delta}}$ can be increased until

$$P_{O_{2,\delta}} = \frac{K_{ev}\sqrt{RT}\,P_{Fe}{}^*\delta}{2D_{O_2,g}} \tag{6-3-32}$$

at which point Y=0 and the FeO coats the surface of the iron. Thus, the loss of iron is sharply reduced.

The maximum rate of iron loss can be determined by substituting eq. (6-3-32) into eq. (6-3-31).

$$\left(\frac{\dot{n}_{Fe}}{A}\right)_{ev} = K_{ev}\left\{\frac{\dfrac{2D_{O_2,g}K_{ev}\sqrt{RT}\,P_{Fe}^*\delta}{2D_{O_2,g}} + D_{Fe,g}P_{Fe}^*}{\sqrt{RT}\,D_{Fe,g} + \delta RT\,K_{ev}}\right\} \tag{6-3-33}$$

(6-3-31)式からは図6-3-1のように蒸発速度は酸素分圧$P_{O_{2,Y}}$の一次の関数であることがわかる。

さらに(6-3-28)式についてみると、$P_{O_{2,\delta}}$は(6-3-32)式の値までは増加し続けるが、Y=0においてFeOが溶鋼表面を覆うことになるため、鉄の蒸発速度、すなわち鉄ロス速度の急激な低下が発生すると理解される。

鉄ロスの最大値は、(6-3-32)式を(6-3-31)式に代入する事によって得られる(6-3-34)式で与えられる。

$$\left(\frac{\dot{n}_{Fe}}{A}\right)_{ev} = \frac{K_{ev}P_{Fe}^*}{\sqrt{RT}} \tag{6-3-34}$$

なお、この式は自由蒸発について適用されることに留意されたい。

This equation applies simply to free evaporation.

Problem 4
The Electrowinning of Zinc

The production of zinc by the sequence of roasting, leaching and electrowinning has won considerable favor compared to other methods involving pyrometallurgy and distillation. In a typical operation, the zinc ore (a complex sulphide) is roasted to convert zinc sulphide to oxide which then is soluble in sulphuric acid. The calcine from the roasters is leached in acid and neutral leaches by spent electrolyte (containing about 110 gpl H_2SO_4) that has been recirculated from the electrowinnig plant. At this stage, both C company and H company also leach zinc fume produced by the zinc fuming furnaces in the lead smelter. The resulting leach liquor with a zinc concentration of approximately 130 gpl is the feed to the electrowinning cells.

The electrowinning cells normally consist of an inert lead or lead-1% silver anode and an aluminum starting sheet for the cathode. Typically there may be 25 anodes and 24 cathodes per deposition tank; the zinc bearing electrolyte is circulated from tank to tank in cascades. With the cells operating at a voltage of 3.6V, zinc is

亜鉛の焙焼、溶解、さらに電解析出と続く湿式精錬工程は、高温プロセスや蒸留プロセスに比べて、比較的容易であるという点で関心を集めている。典型的な湿式精錬では亜鉛鉱石（複雑な組成を持つ硫化鉱）は酸化亜鉛に焙焼され、続いて硫酸に溶解される。ロースターで煆焼された後、電解採取工場から循環される110g/ℓの濃度の硫酸を含む電解質溶液中に浸出（溶解）される。C社やH社では鉛精錬工程で生成した酸化亜鉛についても同時に処理している。このようにして製造した約130g/ℓの濃度の浸出溶液が電解採取装置に供給される。

電解槽は鉛陽極と初期陰極としてのアルミニウム電極から成っている。一般に、電解析出槽には25枚の陽極と24枚の陰極が設置され、亜鉛を含む電解液はこの槽の中を循環している。3.6Vの電圧によって亜鉛は陰極に析出し酸素が陽極から発生する。24時間後に約25ポンド（約11.3kg）の亜鉛が析出した陰極が取り出される。

H社では陽極と陰極は約3cmの間隔で設置され、平均54g/ℓの亜鉛濃度の電解液が図6-4-1に示すように電極間を0.35ガロン/分（英ガロンを4.546ℓとして1.59ℓ/分）の流量で流れている。

陰極前面の拡散層の厚みδは(6-4-1)式で与えられるとする。

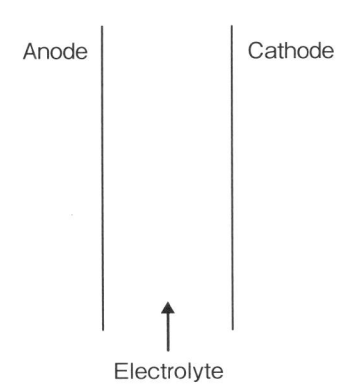

Fig. 6-4-1 Plan view of zinc electrowinning cell.

deposited at the cathode, and oxygen gas is formed at the anode. The cathodes are removed after twenty four hours by which time about 25 lb of zinc per cathode has deposited.

At H company the anodes are separated from the cathodes by a gap of about 3 cm. Electrolyte with an average zinc concentrarion of 54 gpl runs between the electrodes (vide Fig.6-4-1) at a rate of 0.35 gal. per minute.

If the thickness, δ, of the diffusion layer, next to the cathode can be calculated from the relation

$$\delta = \frac{3L^{1/2}}{u_b^{1/2}} \nu^{1/6} D_{Zn^{2+}}^{1/3} \qquad (6\text{-}4\text{-}1)$$

where L is the distance from the edge of the electrode in the direction of liquid flow (cm)

u_b is the bulk velocity of the electrolyte parallel to the electrode (cm sec^{-1})

ν is the kinematic viscosity of the electrolyte (cm^2 sec^{-1})

$D_{Zn^{2+}}$ is the diffusivity of the zinc ion in the electrolyte (cm^2 sec^{-1})

What is the maximum current density* at which you would expect the cell to operate? The following data are available:

Density of the electrolyte = 1.33 g cm^{-3}
Viscosity of the electrolyte = 1.2 cP
Diffusivity of zinc ion in electrolyte
 = 0.72(10^{-5}) cm^2 sec^{-1}
Width of cathode, L, = 61 cm
Height of cathode = 79 cm

ここでLは流動方向の電極の長さ（幅）、u_bは電極幅方向の電解質の流速、νは電解質液の動粘度、$D_{Zn^{2+}}$は電解質液中の亜鉛イオンの拡散係数である。

以上の条件、および右記のデータを基に、この電解析出槽を稼働させるために必要な最大電流密度を推算せよ。なお最大電流密度は限界電流密度とも呼ばれる。

Molecular weight of zinc = 65.4 g mole^{-1}

Faraday's constant = 96,500 C mole^{-1}

If it is necessary to increase the value of the maximum current density, how do you improve the operation conditions?

この最大電流密度を増加させる必要がある場合、操業条件をどのように改善すればよいか提案せよ。

*This term is often referred to as the "limiting current density".

Problem 4 Solution

電解質溶液中の亜鉛イオンは陰極表面に移動し析出する。(i)析出組織への亜鉛の組み込みや(ii)(6-4-2)式で示される陰極反応の電子の移動が速い場合は、バルクから陰極表面への亜鉛イオンの移動が律速段階になる。

特に後者についてはカソードにおける電流密度を制限する過程となる。そこでの物質移動の式(亜鉛析出面への亜鉛イオン流束の式)は(6-4-3)式で表される。

ここで$k_{Zn^{2+}}$は亜鉛イオンの物質移動係数、$C_{Zn^{2+}b}$, $C_{Zn^{2+}int}$はそれぞれバルクおよび析出界面の亜鉛イオン濃度である。

この移動課程を図6-4-2に模式的に示す。

At the cathode, zinc ions in solution must transfer from the bulk of the electrolyte to the electrode surface where they are deposited. If processes such as (i) incorporation of zinc into the matrix of the deposit and (ii) electron transfer in the cathode reaction

$$Zn^{2+} + 2e = Zn \qquad (6\text{-}4\text{-}2)$$

are fast, then the transport of zinc ion from the bulk of the electrolyte to the cathode surface will be the "rate controlling step". Clearly this latter process will also place a limit on the current density we can expect at the cathode. We use our mass transfer equation to write

$$\frac{\dot{n}_{Zn}}{A} = k_{Zn^{2+}}(C_{Zn^{2+}b} - C_{Zn^{2+}int}) \qquad (6\text{-}4\text{-}3)$$

where $\frac{\dot{n}_{Zn}}{A}$ is the flux of zinc ions to the interface (liquid solid) (mole cm^{-2} sec^{-1})

$k_{Zn^{2+}}$ is the mass transfer coefficient of Zn^{2+} in the electrolyte (cm sec^{-1})

$C_{Zn^{2+}b}$, $C_{Zn^{2+}int}$ are the molar concentrations of Zn^{2+} respectively in the bulk of the electrolyte and at the electrolyte-cathode interface (mole cm^{-3}).

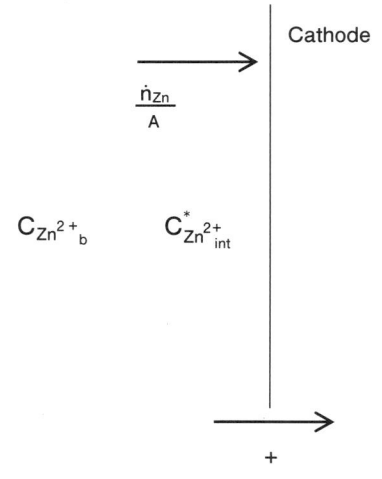

Fig. 6-4-2 Pictorial description of Zn ion transfer process in the vicinity of cathode.

The transfer process is shown pictorially in Fig. 6-4-2.

One can see from our flux equation that for a given $C_{Zn^{2+}_b}$ and $k_{Zn^{2+}}$ the maximum flux of Zn^{2+} will occur when $C_{Zn^{2+}_{int}}=0$. Then

$$\frac{\dot{n}_{Zn^{2+}}}{A}=k_{Zn^{2+}}C_{Zn^{2+}_b} \qquad (6\text{-}4\text{-}4)$$

This equation gives us also our "limiting current density".

Our calculation now involves the evaluation of each of the terms in the equation

[i] $C_{Zn^{2+}_b}$ is given as 54 gpl, and is constant with time.

In terms of $\frac{\text{mole}}{\text{cm}^3}$ the $C_{Zn^{2+}_b}$ is $54\frac{g}{\ell}\times\frac{\text{mole}}{65.4g}\times\frac{1}{1000\text{cm}^3}=8.26(10^{-4})\frac{\text{mole}}{\text{cm}^3}$

[ii] $k_{Zn^{2+}}$ can be evaluated from the film theory. One can picture a diffusion layer existing next to the cathode as the electrolyte flows by. From the film theory we know that

$$k_{Zn^{2+}}=\frac{D_{Zn^{2+},s}}{\delta} \qquad (6\text{-}4\text{-}5)$$

where $\quad \delta=\dfrac{3L^{\frac{1}{2}}\nu^{\frac{1}{6}}(D_{Zn^{2+},s})^{\frac{1}{3}}}{U_b^{\frac{1}{2}}} \qquad (6\text{-}4\text{-}6)$

Substituting back into the flux equation

$$\frac{\dot{n}_{Zn^{2+}}}{A}=\frac{D_{Zn^{2+},s}}{\dfrac{3L^{\frac{1}{2}}\nu^{\frac{1}{6}}(D_{Zn^{2+},s})^{\frac{1}{3}}}{U_b^{\frac{1}{2}}}}\cdot C_{Zn^{2+}_b} \qquad (6\text{-}4\text{-}7)$$

This gives us the flux as a function of distance downstream from the cathode's leading edge, L.

The total rate of zinc transfer is

$$\dot{n}_{Zn^{2+}}=\int_0^{L_o}\frac{D_{Zn^{2+},s}^{\frac{2}{3}}}{3L^{\frac{1}{2}}\nu^{\frac{1}{6}}}U_b^{\frac{1}{2}}\cdot C_{Zn^{2+}_b}H\,dL \qquad (6\text{-}4\text{-}8)$$

where H and L_0 are cathode height and width respec-

(6-4-3)式より、最大のモル流束は $C_{Zn^{2+}_{int}}=0$ の時である事は明らかで(6-4-4)式で与えられる。この式は同時に限界電流密度を与える式でもある。

まず、(6-4-4)式を構成する諸量について求めていく。

[i] $C_{Zn^{2+}_b}$

単位を変換するとすると次のようになる；$8.26(10^{-4})\frac{\text{mole}}{\text{cm}^3}$

[ii] $k_{Zn^{2+}}$

境膜理論から(6-4-5)式のようになる。ここで δ は(6-4-6)式で与えられる。

(6-4-5)、(6-4-6)式を(6-4-4)式に代入して(6-4-7)式を得る。

(6-4-7)式を電極の幅方向に積分すると全モル移動速度は(6-4-9)式のようになる。

tively.

$$\dot{n}_{Zn^{2+}} = \frac{2}{3} D_{Zn^{2+},s}^{\frac{2}{3}} \frac{u_b^{\frac{1}{2}}}{\nu^{\frac{1}{6}}} \cdot C_{Zn^{2+}_b} L_0^{\frac{1}{2}} H \qquad (6\text{-}4\text{-}9)$$

したがって平均モル流束は(6-4-10)式で与えられる。

The average zinc ion flux is obtained by dividing the equation for $\dot{n}_{Zn^{2+}}$ by the area of the cathode, $A = HL_0$. Then

$$\frac{\dot{n}_{Zn^{2+}}}{A} = \frac{2}{3} \frac{D_{Zn^{2+},s}^{\frac{2}{3}} u_b^{\frac{1}{2}} C_{Zn^{2+}_b}}{\nu^{\frac{1}{6}} L_0^{\frac{1}{2}}} \qquad (6\text{-}4\text{-}10)$$

(a) $u_b^{\frac{1}{2}}$、(b) $L_0^{\frac{1}{2}}$、(c) $D_{Zn^{2+}}^{\frac{2}{3}}$、(d) $\nu^{\frac{1}{6}}$ をそれぞれ求めると左記のようになり、

$$\frac{\dot{n}_{Zn^{2+}}}{A} = 1.9256 \times 10^{-8} \left(\frac{\text{mole}}{\text{cm}^2 \cdot \text{s}} \right)$$

と求められる。

(a) Calculate $u_b^{\frac{1}{2}}$

$$u_b = \frac{\text{volume flow rate of electrolyte}}{\text{X-sectional area of channel}}$$

$$= \frac{0.35 \frac{\text{gal}}{\text{min}} \times \frac{\text{min}}{60 \text{sec}} \times \frac{4.55(10^3) \text{cm}^3}{\text{gal}}}{79 \text{cm} \times 3 \text{cm}}$$

$$= 0.112 \frac{\text{cm}}{\text{sec}}$$

and

$$u_b^{\frac{1}{2}} = 0.335 \frac{\text{cm}^{\frac{1}{2}}}{\text{sec}^{\frac{1}{2}}}$$

(b) Calculate $L_0^{\frac{1}{2}}$

$$L_0^{\frac{1}{2}} = \sqrt{61}$$

$$= 7.81 \text{cm}^{\frac{1}{2}}$$

(c) Calculate $D_{Zn^{2+}}^{\frac{2}{3}}$

$$D_{Zn^{2+}}^{\frac{2}{3}} = (7.2(10^{-6}))^{\frac{2}{3}}$$

$$= 3.72(10^{-4}) \frac{\text{cm}^{\frac{4}{3}}}{\text{sec}^{\frac{2}{3}}}$$

(d) Calculate $\nu^{\frac{1}{6}}$

$$\nu^{\frac{1}{6}} = \left(\frac{0.012}{1.33} \right)^{\frac{1}{6}}$$

$$= 0.4563 \frac{\text{cm}^{\frac{1}{3}}}{\text{s}^{\frac{1}{6}}}$$

from (a),(b),(c),(d) and [i],

$$\frac{\dot{n}_{Zn^{2+}}}{A} = \frac{2}{3} \frac{(3.72 \times 10^{-4}) \times (0.335) \times (8.26 \times 10^{-4})}{(0.4563) \times (7.81)}$$

$$= 1.9256 \times 10^{-8} \left(\frac{mole}{cm^2 \cdot sec}\right)$$

Finally the current density $\left(\frac{i}{A}\right)$ is calculated

$$\frac{i}{A} = 2F\left(\frac{\dot{n}_{Zn^{2+}}}{A}\right) = 2 \times 96500 \times 1.9256 \times 10^{-8}$$

$$= 3.716 \times 10^{-3} (A \cdot cm^{-2})$$

One of the ways to increase the value of the current density efficiently is the increase of flow velocity of the electrolyte parallel to the electrode.

The flow velocity in operation estimated from the given conditions seems to be very low (0.1cm/sec). If it could be increased to 10cm/sec, the maximum current density will be 10 times higher than the present value.

最終的に電流密度 $\frac{i}{A}$ は $2F\left(\frac{\dot{n}_{Zn^{2+}}}{A}\right)$ で与えられるため（Fはファラデー定数）、限界電流密度は $3.716 \times 10^{-3} (A \cdot cm^{-2})$ となる。

効率的に電流密度を上げるには、電極に平行な向きの電解質溶液の流速を増加させることが提案される。ここで示される操業条件から推定される流速は0.1cm/秒と非常に低い値である。もしこの流速を10cm/秒に増加することができれば、最大電流密度は現状の10倍になると期待される。

Problem 5
The Dissolution Rate of Tungsten Pellets Falling Through the Liquid Pool of a Continuously Cast Steel Billet

鋼の連続鋳造を首尾よく実施する上で鋳型から凝固を開始し、ある長さにわたって溶鋼を保持するシェルの特性を把握する事は非常に重要である。たとえば、鋳型下でのシェル厚の不均一部位や割れはシェルの破断やブレークアウト（シェルの割れた部位からの溶鋼の噴出）につながる。

W社で凝固シェルの厚みや溶鋼プール深さに及ぼすプロセス変数の影響を調査するためのプログラムがスタートした。

プール深さの測定は2つの球状タングステンペレットを用いて行われる。これらペレットは少量のCo^{60}をタングステン（融点3377℃）で包んだ構造となっている。

図6-5-1に示すように、鋳造途中で、一方のペレットを鋳型内メニスカスレベルの凝固シェルに付着させ凝固開始位置とすると共に、他方のペレットを鋳型メニスカス中央から落下させてプールの底の位置をマーキングする。

その後、ガイガーカウンターによって、2つのペレットの位置および間隔を測定する。

The successful continuous casting of steel depends largely on the ability of the solid steel shell, which has solidified in the continuous casting mold, to support the liquid pool for some distance down the strand. A thin or cracked section of the shell below the mold may lead to rupture of the casting and breakout of the molten iron.

A program has been initiated at W company to study the variables which affect the thickness of the solid shell and the depth of the liquid pool. The pool depth is measured with the aid of two spherical tungsten* pellets in which small quantities of radioactive Co^{60} are sealed. During casting, one pellet is inserted into a corner of the mold at meniscus level, where it is frozen into the solid shell.

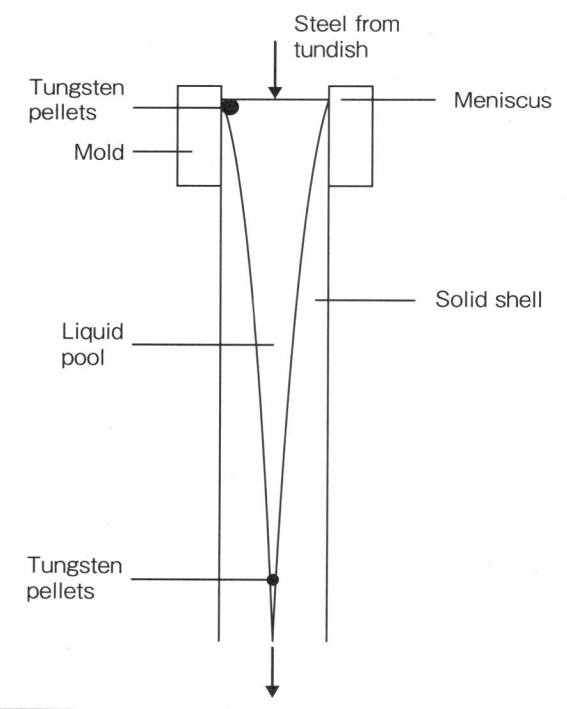

Fig. 6-5-1　Pool depth determination in billet casting.

*Tungsten has a melting point of 3377℃.

Simultaneously, the bottom of the pool is marked by dropping the second pellet into center of the mold, and through the liquid steel (vide Fig. 6-5-1). Later with geiger counters, the pellets location in the billet and the distance between them is measured.

Because of its long half-life, the possibility of Co^{60} escaping into the steel must be seriously considered. Such an event is not likely with the pellet at the meniscus, but is possible with the second tungsten pellet to dissolve completely during its fall through the liquid steel.

Check on this latter possibility by calculating the quantity of tungsten that is dissolved during the pellet's descent.

その長い半減期のため、前もってCo^{60}の溶鋼への溶解が生じないことを厳密にチェックする必要がある。この問題はメニスカスのペレットでは発生しないと考えられるが、溶鋼プール中を落下するペレットではタングステンが溶鋼中に溶け出すことも起こりうると危惧される。
この後者の事象が起こる可能性を、ペレット降下中の溶鋼のタングステン溶解量を計算して推定せよ。

A reasonable estimate of the liquid phase mass transfer coefficient around a sphere is frequently obtained by applying the penetration theory; one then uses the time taken for the sphere to fall through its diameter as the surface renewal time.

The terminal descent velocity of the pellet through the steel, u_t, may be estimated from the relation

$$u_t = 1.73 \sqrt{\frac{d_w \ g(\rho_w - \rho_{Fe})}{\rho_{Fe}}} \qquad (6-5-1)$$

液相での固体球周囲の物質移動係数の推算には、浸透説が適用できる。球体がその直径の距離を落下する時間をもって表面更新時間とし、球体の落下速度は(6-5-1)式が適用できるものとする。

where d_w is the diameter of the tungsten pellet (cm)
   g is gravitational acceleration (980 cm sec^{-2})
   ρ_w is the density of tungsten (g cm^{-3})
   ρ_{Fe} is the density of molten steel (g cm^{-3}).

The phase diagram for tungsten-iron is given in Fig.6-5-2.
  Temperature of the liquid pool** = 1550℃
  Length of the liquid pool   = 13 ft
  Density of tungsten   = 19 g cm^{-3}

計算に使用する諸量は左記のとおりである。

Density of liquid steel = 7.2 g cm^{-3}
Molecular weight of tungsten = 183.9 g mole^{-1}
Molecular weight of iron = 55.85 g mole^{-1}
Diameter of the tungsten pellet = 2.0 cm
Diffusion coefficient of tungsten in steel
= 4(10^{-5}) cm^2 sec^{-1}

**The pool temperature is rarely this high, but will be used here as an extreme case.

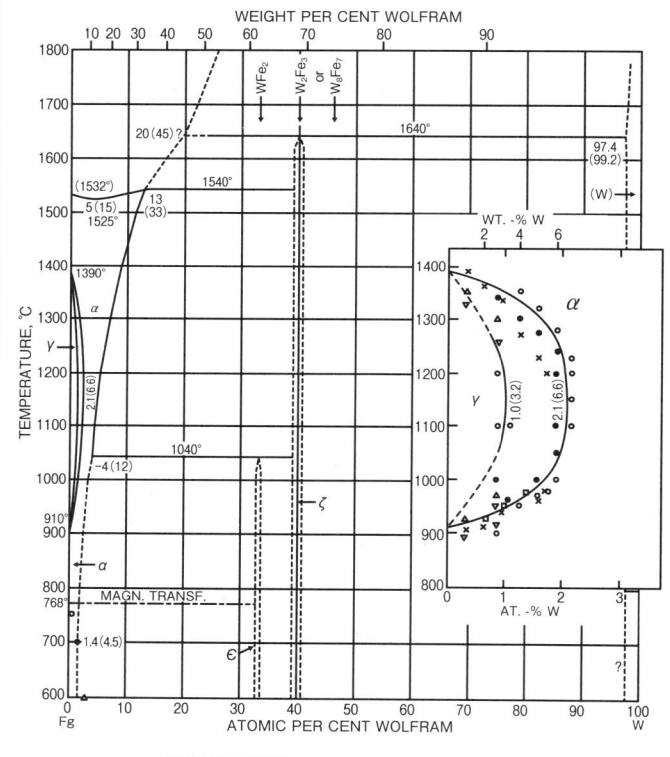

Fig. 6-5-2 Fe-W phase diagram.

Problem 5 Solution

Upon examination it would seem clear that the rate of dissolution of tungsten would not depend on any diffusional processes inside the pellet, since it is pure tungsten.

ペレットが純タングステンであるがゆえに、このペレットの溶解速度にはペレット内部の拡散は関与していないと考える。

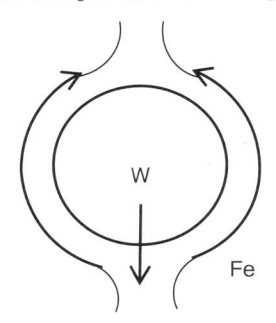

Fig. 6-5-3 Tungsten falling down molten steel pool.

The rate of dissolution could depend, therefore on two other processes:
(i) the actual dissolution whereby at the interface tungsten atoms are removed from their crystal lattice to go into the liquid iron
(ii) the mass transfer of dissolved tungsten from the solid-liquid interface into the bulk of the liquid iron.

ペレットの溶解速度には以下の2つのプロセス(i)、(ii)の可能性を考える。

(i) 界面においてタングステン原子が結晶格子から離れ、直接溶鋼中に移行する。

(ii) 固液界面において、溶解したタングステンが界面から溶鋼バルクへ移動する。

Let us assume, that at the high temperatures encountered here, step (i) is very fast, so that (ii) mass transfer of tungsten is the "rate controlling" step.

本系のような高温では(i)の段階は非常に速く進行するため、(ii)のタングステンの界面での移動が律速段階となると考えられる。

Using the phenomenological equation, we may express the rate of mass transfer of tungsten away from the pellet in the following way.

ペレットからのタングステンの移動速度は(6-5-2)式で記述される。

$$\frac{\dot{n}_W}{A} = k_W(C_{W_{int}} - C_{W_b}) \qquad (6-5-2)$$

ここで、k_Wは溶鋼中のタングステンの物質移動係数、$C_{W_{int}}$はペレット／溶鋼界面における溶鋼中のタングステン濃度、C_{W_b}はバルク溶鋼中のタングステン濃度である。

界面でのタングステンの移動が律速過程との仮定の下、$\dot{n}_W$はタングステンの溶解速度となる。

次に(6-5-2)式で用いた数値について評価する。

(a) ペレットに比べ溶鋼量は十分に大きいため、$C_{W_b} \simeq 0$とみなすことができる。
(b) k_Wは浸透説から計算される。
(c) $C_{W_{int}}$はW-Fe状態図から求めることができる。鉄とタングステンは界面で平衡であり界面での溶解速度はきわめて速いと考える。

図6-5-2よりペレット/溶鋼界面のタングステン濃度は33wt%と推定される。

wt%からモル濃度への変換は左記のようになり、$C_{W_{int}}$は1.62×10^{-2}mole/cm^{3}となる。

where $\dfrac{\dot{n}_W}{A}$ is the flux of tungsten away from the pellet surface (mole cm^{-2} sec^{-1})

k_W is the mass transfer coefficient of tungsten in the liquid iron (cm sec^{-1})

$C_{W_{int}}$ is the concentration of tungsten in the liquid iron at the pellet iron interface (mole cm^{-3}).

C_{W_b} is the concentration of tungsten in the bulk of the liquid iron (mole cm^{-3})

Note here that $\dot{n}_W$ is actually the rate of dissolution of tungsten since mass transfer is the rate controlling step. Let us evaluate some of the terms in the equation above.

(a) $C_{W_b} \simeq 0$ since tha volume of the steel is so much greater than that of the pellet.
(b) k_W can be calculated using the penetration theory.
(c) $C_{W_{int}}$ can be calculated from the W-Fe phase diagram, since the Fe and W at the interface will be at equilibrium (the dissolution step at the interface is very fast).

From Fig. 6-5-2 the concentration of W at the interface is estimated to be 33 wt % W. This should be converted to mole/cm^3 for the equation. Let us calculate the volume of 100 g of 33% W-Fe. Assume can add individual vols. such that

$$\frac{33}{\left(\frac{19\text{g}}{\text{cm}^3}\right)} + \frac{67}{\left(\frac{7.2\text{g}}{\text{cm}^3}\right)} = 11.04 \text{cm}^3$$

Therefore the molar concn. is

$$\frac{\frac{33}{184}\text{mole}}{11.04\text{cm}^3} = 1.62(10^{-2})\frac{\text{mole}}{\text{cm}^3}$$

(d) $\dot{n}_W$, from a balance on W in the sphere, is

$$\dot{n}_W = -C_{W_S} \cdot 4\pi r^2 \frac{dr}{dt} \qquad (6\text{-}5\text{-}3)$$

where C_{W_S} is the molar concentration of solid tungsten, r is the pellet radius.

$$\therefore \frac{\dot{n}_W}{A} = -C_{W_S} \frac{dr}{dt} \qquad \text{because } A = 4\pi r^2 \qquad (6\text{-}5\text{-}4)$$

From eq.(6-5-2) and eq. (6-5-4)

$$\frac{dr}{dt} = -\frac{k_W}{C_{W_S}} C_{W_{int}}$$

$$k_W = 2\sqrt{\frac{D_{W,Fe}}{\pi t_e}} \qquad (6\text{-}5\text{-}5)$$

where $t_e = \frac{d_W}{u_t}$, and $D_{W,Fe}$ is the diffusivity of W in Fe.

$$\therefore k_W = 2\sqrt{\frac{D_{W,Fe} u_t}{\pi d_W}} \qquad (6\text{-}5\text{-}6)$$

u_t is estimated from the given relation

$$u_t = 1.73\sqrt{\frac{d_W g(\rho_W - \rho_{Fe})}{\rho_{Fe}}}$$

We will assume in our calculation, that as a first approximation d_W is constant (approximately).

$$\therefore u_t = 1.73\sqrt{\frac{2.0\,\text{cm} \cdot 980\,\frac{\text{cm}}{\text{sec}^2}(19-7.2)\,\frac{\text{g}}{\text{cm}^3}}{\frac{7.2\,\text{g}}{\text{cm}^3}}}$$

$$= 98.1\,\text{cm sec}^{-1}$$

$$\therefore k_W = 2\sqrt{\frac{4(10^{-5})\frac{\text{cm}^2}{\text{sec}} \cdot 9.81(10^1)\frac{\text{cm}}{\text{sec}}}{\pi \cdot 2.0\,\text{cm}}}$$

$$= 5.0(10^{-2})\,\text{cm sec}^{-1}$$

$$\int_{r_o}^{r} dr = -\frac{k_W}{C_{W_S}} C_{W_{int}} \int_{0}^{t} dt \qquad (6\text{-}5\text{-}7)$$

Here t of interest is the time taken for the pellet to fall through the liquid pool, i.e.

$$t = \frac{13\,\text{ft} \times 12\,\frac{\text{in}}{\text{ft}} \times 2.54\,\frac{\text{cm}}{\text{in}}}{98.1\,\frac{\text{cm}}{\text{sec}}}$$

(d) 球状タングステンの溶解速度（半径の減少速度）は(6-5-3)式で表される。

(6-5-2)式と(6-5-3)式より(6-5-5)式が得られる。

(6-5-5)式をr、tで積分すると(6-5-7)式、(6-5-8)式のようになる。

$$= 4.04 \text{ sec.}$$

$$r - r_o = -\frac{k_W}{C_{W_S}} C_{W_{int}} t \tag{6-5-8}$$

$$= \frac{-5(10^{-2})\frac{\text{cm}}{\text{sec}} \cdot 1.62(10^{-2})\frac{\text{mole}}{\text{cm}^3} \cdot 4.04 \text{sec}}{\dfrac{19\frac{\text{g}}{\text{cm}^3}}{184\frac{\text{g}}{\text{mole}}}}$$

$$= -3.16(10^{-2}) \text{cm}.$$

∴ $r \simeq r_0$ and it would appear that very little dissolution of the pellet will occur.

最終的に計算されるようにrの減少は3.16×10^{-2}cmとわずかであり、ペレットの溶解はほとんど無視できる。

Problem 6
The Prediction of Decarburization Rates in the Molten Steel Experiment

In our laboratory experiment a shallow bath of molten iron is in contact with a slag containing a high concentration of FeO. The high oxidizing power of the FeO in the slag allows the selective removal from the iron of dissolved impurities such as silicon, manganese, carbon and phosphorus. The FeO combines with the silicon, manganese, and phosphorus to form oxides which are insoluble in the metal, but soluble in the slag. Carbon is oxidized to CO gas, bubbles of which nucleate on the refractory lining. This process is the genesis of the "carbon boil" since the CO bubbles grow, break away from their nucleation sites and rise through the metal and slag which contributes the promotion of slag/metal reaction.

浅く広い溶銑プールとその上の高濃度のFeOを含有する溶融スラグから成る系における溶銑の脱炭反応につい考察する。

この酸化能力の高いスラグは溶銑中の不純物であるSi、Mn、C、およびPの選択的除去を可能にする。酸化されるC以外の元素はスラグに溶解して除去されるが、CはCOガスとして炉のライニング表面で核生成してメタル、スラグ中を上昇して系外に除去される。CO気泡が成長しスラグ／メタル界面を撹拌する現象は"カーボンボイル"と呼ばれ、反応速度の促進に貢献すると期待されている。

The carbon-oxygen reaction may be visualized as occurring through a series of steps,

(1) transfer of FeO from the bulk of the slag to the slag-metal interface
(2) reaction at the interface
$$FeO_{(slag)} = Fe + O_{(Fe)} \qquad (6\text{-}6\text{-}1)$$
(3) transfer of dissolved oxygen from the slag-metal interface to the bulk of the metal.
(4) transfer of (a) oxygen and (b) carbon, dissolved in the metal, from the bulk metal to the metal – CO bubble interface
(5) reaction at the metal-bubble interface
$$C_{(Fe)} + O_{(Fe)} = CO_{(g)} \qquad (6\text{-}6\text{-}2)$$
(6) rise of CO bubbles through metal and slag into the upper furnace atmosphere (carbon boil).

These steps are shown schematically in Fig.6-6-1.

このC-O反応は以下のステップで進行するものと考えられる。

(1) バルクスラグからのFeOのスラグ／メタル界面への移動
(2) (6-6-1)式で示されるスラグ／メタル界面での反応
(3) スラグ／メタル界面からバルク溶融メタル中への酸素の移動
(4) 溶融メタル中の酸素、CのCO気泡表面への移動
(5) (6-6-2)式で示されるCO気泡発生反応
(6) CO気泡の発生サイトからの離脱、溶融メタル、スラグ中の浮上とガス雰囲気中への移動

これら過程を図6-6-1に模式的に示す。

図6-6-1においてステップ(2)、(4b)、(5)、(6)の速度は非常に速く律速過程とはなりえないと仮定し、脱炭反応速度$\dot{n}_c$を記述する式を導け。なお、この式の導出にあたっては(1)、(3)、(4a)の各酸素の移動段階を考慮すると共に、これらのステップが総括脱炭速度にどのように関係しているかを示せ。

In considering the rate at which the carbon boil occurs, it has been argued that steps (2), (4b), (5), and (6) are sufficiently fast that they are not rate controlling. Assuming that this is the case, derive an expression for the rate of decarburization of iron, $\dot{n}_c$. Note that this expression should take into consideration all three of the oxygen transport steps. Show mathematically the conditions under which each step would control the overall rate of decarburization.

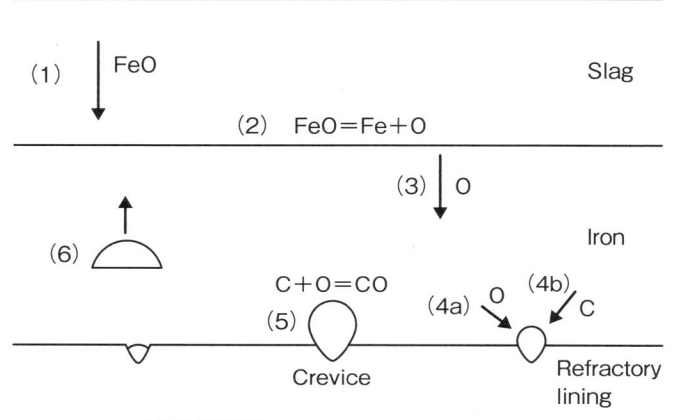

Fig. 6-6-1 Steps in the carbon boil.

最後に、左記に示す条件に基づき溶銑浴の脱炭速度を計算せよ。

With the data given below for a typical experimental coditions, calculate the rate of decarburization of the iron bath.

Temperature of iron and slag = 1600℃
Equilibrium constant for step (2) at 1600℃
$$= 4.5 \, (10^{-3}) \frac{\text{wt\%O}}{\text{wt\%FeO}}$$
Equilibrium constant for step (5) at 1600℃
$$= 500 \text{ atm. wt\%}^{-2}$$
Bulk concentration of FeO in the slag = 20wt%
Bulk concentration of oxygen in the iron = 0.1 wt%

Bulk concentration of carbon in the iron = 0.5 wt%

Density of the slag = 3.5 g cm^{-3}

Density of the iron = 7.0 g cm^{-3}

Molecular weight of iron = 55.85 g mole^{-1}

Molecular weight of oxygen = 16.0 g mole^{-1}

Molecular weight of carbon = 12.01 g mole^{-1}

Metal bath depth = 50 cm

Slag bath depth = 20 cm

Mass transfer coefficient of FeO in the slag
$\qquad$ = 0.003 cm sec^{-1}

Mass transfer coefficient of oxygen in iron at the slag-metal interface = 0.45 cm sec^{-1}

Mass transfer coefficient of oxygen in iron at the metal-bubble interface = 0.05 cm sec^{-1}

Area of slag-metal interface $\simeq 2\ (10^5)$ cm^2

Ratio of area of slag-metal interface to area of bubble-metal interface = 0.5

Compare your answer to the decarburization rate of 0.12 to 0.18 percent carbon per hour obtained in the experiment (see also Fig.6-6-2).

得られた計算結果と実験における脱炭速度（0.12%〜0.18%C/hr）とを比較せよ。

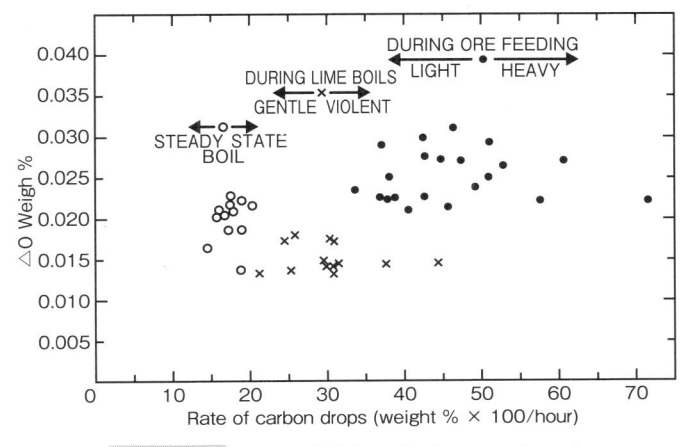

Fig. 6-6-2 Rate of C drop in the experiment.

Problem 6 Solution

図6-6-3にCO気泡発生にかかわるスラグ、メタル中のC濃度分布およびO濃度分布を模式的に示す。

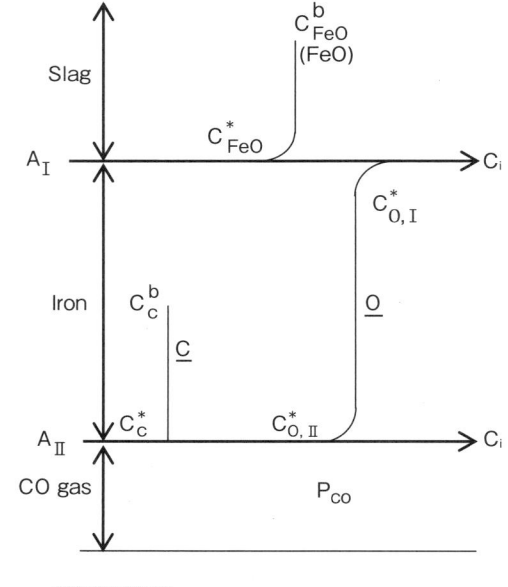

Fig. 6-6-3 Consentration profiles in iron and slag.

<Assumptions>

a) 界面での反応は非常に速い

b) Cの移動は反応律速ではない

c) CO分圧に関しては気泡径の影響を無視する

d) CO気泡の発生は非常に速く、スラグ／メタル界面の移動現象には影響を与えない

a) The reaction takes place instantaneously.

b) Transport of C is not rate controlling. ($C_c^* \simeq C_c^b$)

c) Neglect the effect of bubble diameter on the P_{co}.
$(P_{co} = P_{co}^o + \frac{\alpha\sigma}{d})$

d) CO bubble formed rises quickly and has no effect on the transport phenomena at the slag-metal interface.

仮定a)、b)より、(6-6-3)式が得られる。
さらに仮定c)より気泡発生時のCO分圧、酸素濃度は(6-6-4)、(6-6-5)式のようになる。

From the assumptions a) & b)

$$K_{\mathrm{II}} = \frac{P_{co}}{[\%C]^*[\%O]_{\mathrm{II}}^*} = \frac{P_{co}}{[\%C]_b[\%O]_{\mathrm{II}}^*} \quad (6\text{-}6\text{-}3)$$

From the assumption c)

$$P_{co} = 1.0 + \frac{50 \times 7.0 + 20 \times 3.5}{76 \times 13.6} = 1.41 \text{ (atm)} \quad (6\text{-}6\text{-}4)$$

$$\therefore C_{O,\mathrm{II}}^* = \frac{P_{co} \cdot \rho_{iron} \times 10^{-2}}{K_{\mathrm{II}}[\%C]_b \cdot M_O} = \frac{1.41 \times 7.0 \times 10^{-2}}{500 \times 0.5 \times 16.0}$$

$$= 2.468 \times 10^{-5} \text{ (mole/cm}^3\text{)} \quad (6\text{-}6\text{-}5)$$

$$\frac{\dot{n}_{FeO}}{A_I} = k_{FeO}(C_{FeO}^b - C_{FeO}^*) \tag{6-6-6}$$

$$-\frac{\dot{n}_{O,I}}{A_I} = k_{O,I}(C_O^b - C_{O,I}^*) \tag{6-6-7}$$

$$\frac{\dot{n}_{O,II}}{A_{II}} = k_{O,II}(C_O^b - C_{O,II}^*) \tag{6-6-8}$$

$$K_I' = \frac{C_{O,I}^*}{C_{FeO}^*} \tag{6-6-9}$$

$$\dot{n}_{FeO} = \dot{n}_{O,I} \tag{6-6-10}$$

$$\dot{n}_{FeO} = \dot{n}_{O,II} \tag{6-6-11}$$

スラグメタル界面におけるFeOsとOのモル流束、およびメタル気泡界面におけるOのモル流束はそれぞれ(6-6-6)、(6-6-7)、(6-6-8)式で表される。

スラグメタル界面における平衡は(6-6-9)式のようになる。

化学量論性については、(6-6-10)、(6-6-11)式が得られる。

From (6-6-6), (6-6-7), and (6-6-9)

$$\dot{n}_{O,I} = -A_I k_{O,I}(C_O^b - K_I' C_{FeO}^*)$$
$$= A_I k_{FeO}(C_{FeO}^b - C_{FeO}^*) \tag{6-6-12}$$

From (6-6-6), (6-6-8) and (6-6-9)

$$\dot{n}_{O,II} = A_{II} k_{O,II}(C_O^b - C_{O,II}^*)$$
$$= A_I k_{FeO}(C_{FeO}^b - C_{FeO}^*) \tag{6-6-13}$$

これらの関係から、スラグメタル界面におけるOのモル流束は(6-6-12)式で表され、メタル気泡界面のOのモル流束は(6-6-13)式で表される。

(6-6-12)と(6-6-13)式よりスラグ／メタル界面のFeO濃度が(6-6-14)式のように求まる。

From (6-6-12), and (6-6-13)

$$C_{FeO}^* = \frac{k_{FeO}C_{FeO}^b + k_{O,I}C_O^b}{k_{FeO} + K_I' k_{O,I}} \tag{6-6-14}$$

さらに、(6-6-13)、(6-6-14)式より、メタルバルク中の酸素濃度が(6-6-15)式のように求まる。

From (6-6-12), (6-6-13) and (6-6-14)

$$C_O^b = \frac{1}{\left(\dfrac{1}{A_{II}k_{O,II}} + \dfrac{1}{A_I k_{O,I}} + \dfrac{K_I'}{A_I k_{FeO}}\right)} \left\{ \frac{K_I'}{A_{II}k_{O,II}} C_{FeO}^b + \left(\frac{1}{A_I k_{I,O}} + \frac{K_I'}{A_I k_{FeO}}\right) C_{O,II}^* \right\} \tag{6-6-15}$$

(6-6-8)式と(6-6-15)式よりスラグ／メタル界面を横ぎる酸素のモル流束は(6-6-16)式となり、脱炭反応にかかわる化学量論性(6-6-17)式から脱炭速度は(6-6-18)式のようになる。

From eqs. (6-6-8) and (6-6-15)

$$\frac{\dot{n}_{O,II}}{A_{II}} = \frac{1}{\left(\dfrac{1}{A_{II}k_{O,II}} + \dfrac{1}{A_{I}k_{O,I}} + \dfrac{K'_{I}}{A_{I}k_{FeO}}\right)} \left(\frac{K'_{I}}{A_{II}}C^{b}_{FeO} - \frac{C^{*}_{O,II}}{A_{II}}\right)$$

(6-6-16)

Stoichiometry

$$\frac{\dot{n}_{C}}{A_{II}} = \frac{\dot{n}_{O,II}}{A_{II}} \tag{6-6-17}$$

From eqs. (6-6-16) and (6-6-17)

$$\dot{n}_{C} = \frac{1}{\left(\dfrac{1}{A_{II}k_{O,II}} + \dfrac{1}{A_{I}k_{O,I}} + \dfrac{K'_{I}}{A_{I}k_{FeO}}\right)} (K'_{I} C^{b}_{FeO} - C^{*}_{O,II})$$

(6-6-18)

where $K'_{I} = \dfrac{C^{*}_{O,I}}{C^{*}_{FeO}} = K_{I}\left(\dfrac{\rho_{iron} \cdot M_{FeO}}{\rho_{slag} \cdot M_{O}}\right)$

$\left(K_{I} = \dfrac{[\%O]^{*}}{(\%FeO)^{*}} = 4.5 \times 10^{-3}\right)$

$\rho_{iron} = 7.0 \left(\dfrac{g}{cm^{3}}\right)$

$\rho_{slag} = 3.5 \left(\dfrac{g}{cm^{3}}\right)$

$M_{FeO} = 71.85 \left(\dfrac{g}{mole}\right)$

$M_{O} = 16.0 \left(\dfrac{g}{mole}\right)$

$M_{c} = 12.0 \left(\dfrac{g}{mole}\right)$

$k_{FeO} = 0.003 \left(\dfrac{cm}{sec}\right)$

$k_{O,I} = 0.45 \left(\dfrac{cm}{sec}\right)$

$k_{O,II} = 0.05 \left(\dfrac{cm}{sec}\right)$

$C^{b}_{FeO} = (\%FeO)^{b} \rho_{slag} \times \dfrac{10^{-2}}{M_{FeO}}$

$= 9.743 \times 10^{-3} \left(\dfrac{mole}{cm^{3}}\right)$

$$A_{\text{I}} = 2 \times 10^5 \text{ (cm}^2\text{)}$$

$$A_{\text{II}} = 4 \times 10^5 \text{ (cm}^2\text{)}$$

$$-\frac{d[\%C]}{dt} = \dot{n}_c \times (60)^2 \times \frac{M_c}{L} \times A_{\text{II}} \times \rho_{\text{iron}} \times 10^{-2}$$

$$= \frac{\dfrac{\dot{n}_c \times (60)^2 \times M_c}{L \times A_{\text{I}} \times \rho_{\text{iron}} \times 10^{-2}}}{\left(\dfrac{1}{A_{\text{II}} k_{o,\text{II}}} + \dfrac{1}{A_{\text{I}} k_{O,\text{I}}} + \dfrac{K'_{\text{I}}}{A_{\text{I}} k_{\text{FeO}}}\right)} (K'_{\text{I}} C^b_{\text{FeO}} - C^*_{O,\text{II}})$$

$$= 0.177 \text{ (wt\% C/hr)}$$

(6-6-19)

This calculated value of the decarburization rate is reasonable compared with the measured value (0.12 to 0.18 wt% C/hr).

(6-6-18)式に実験の諸量を代入して計算された脱炭速度の値は、測定された値とよく一致しており、本モデルの前提とした界面における酸素の移動が脱炭反応を律速していると言える。

Problem 7
Oxygen Absorption in an Open-Pour Stream of Contiuous Casting

ビレットやブルームの連続鋳造においては、図6-7-1に示すように溶鋼はタンディッシュから水冷鋳型中に連続的に注入される。

During the continuous casting of steel into semifinished shapes like billets and blooms, steel is poured in a continuous fashion from a tundish into a water-cooled copper mould, as shown in Fig. 6-7-1.

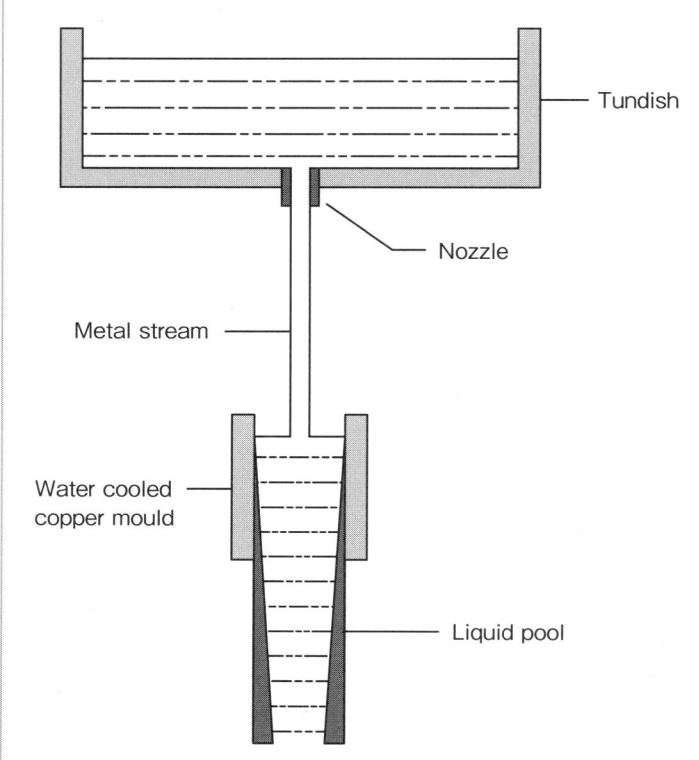

Fig. 6-7-1 Pouring steel from tundish into mould during continuous casting.

要求される鋼の品質により（また、鋳造断面にもより）、オープン注入または浸漬注入が用いられる。オープン注入においては、タンディッシュからの溶鋼の流れは大気からシールされないまま鋳型内の溶鋼プールに供給される。最も簡便な注入プロセスではあるが、この方法では流れは大気に曝されるため酸素と反応し、鋼中の酸素濃度が増加して清浄性を

Depending on the quality of steel desired, either an "open pour" or a "submerged pour" technique is used. With the open pour method, the stream of steel from the tundish is allowed to pass unshielded through the air to the liquid pool below. Although the simplest to operate, this technique has the disadvantage that oxygen can be absorbed from the air by the exposed stream, resulting in an increase in the content of oxide inclusions, and a "dirtier" steel. In the case of submerged pouring, reoxidation of the steel by the air is averted by passing the steel through a refractory tube which is attached to the tundish at one end, and immersed in the liquid pool at the other end.

At L company the open pour method is utilized. In this operation, steel is poured from a tundish at a rate to 5.4 kg/sec through a zirconia nozzle with an internal diameter of 2.22 cm. If the oxygen content of the steel in the tundish is 4.5 (10^{-6}) mole cm^{-3}, and the distance between the tundish and the surface of the liquid pool in the mould is roughly 45 cm, calculate the bulk concentration of oxygen in the steel at the liquid pool level. The equilibrium constant, K', for the absorption reaction

$$1/2\ O_2 = [O]_{Fe} \quad (6\text{-}7\text{-}1)$$

is 1.60 mole cm^{-3} $atm^{-1/2}$ at the steel temperature of 1540℃. Other available data are listed below:

Density of molten steel @ 1540℃ = 7.4 g cm^{-3}
Diffusion coefficient of oxygen in air = 1.0 cm^2 sec^{-1}
Mean air temperature = 770℃
Partial pressure of oxygen in air = 0.21 atm
Gas constant, R = 82.1 cm^3 atm $mole^{-1} K^{-1}$

The following assumptions may be made:
[ⅰ] the absorption reaction is instantaneous.
[ⅱ] mass transfer within the highly turbulent liquid phase is sufficiently fast not to be rate controlling so that bulk and interfacial concentrations in the steel are about equal.
[ⅲ] the penetration theory is applicable.

損なう欠点がある。一方、浸漬法では、溶鋼は鋳型内溶鋼プールとをつなぐ耐火物を通してタンディッシュから鋳型内に供給されるため溶鋼は再酸化されにくい。

L社はオープン注入で操業を行っている。この操業では溶鋼はタンディッシュの底に設けた内径2.22cmのジルコニアノズルを通して5.4 kg/秒の流量で鋳型内に注入される。タンディッシュ内溶鋼の酸素含有量が4.5 (10^{-6}) mole cm^{-3}、タンディッシュ出口とプール表面との距離が45cmである場合、鋳型内メニスカス（溶鋼プール表面）における溶鋼中の溶融酸素濃度を求めよ。

気体中酸素が溶鋼中に溶解する反応は(6-7-1)式で表され、その平衡定数は1540℃において1.60 mole cm^{-3} $atm^{-1/2}$である。
他の諸量は左記の通りである。

計算に当たっては次の仮定を前提とせよ。
[ⅰ] 溶鋼の酸素吸収反応は極めて速い。
[ⅱ] 溶鋼流れは乱流状態にあり、溶鋼中の酸素の移動は律速とはならない。すなわち溶鋼流れのバルクと大気との界面における酸素濃度は等しい。
[ⅲ] 溶鋼/大気界面での酸素吸収反応には、浸透説を適用する。

[iv] タンディッシュ底部から鋳型内溶鋼プールへの注入流形状は円柱（形状一定）とみなす。
[v] 注入流中の溶鋼は半径方向に完全混合であり、落下する方向にはプラグフロー（押出し流れ）である。
[vi] 注入流中の溶鋼温度は一定である。
[vii] 溶鋼中に吸収された酸素は酸素原子として溶解しており、溶鋼中の他の元素とは反応しない（酸化物を形成しない）。
[viii] 系は定常状態にあると仮定する。

[iv] the stream of metal retains its cylindrical shape between tundish and liquid pool.
[v] the stream is well mixed radially, but falls in plug flow.
[vi] heat losses from the stream are negligible.
[vii] oxygen that is absorbed from the air remains dissolved, and does not combine with other dissolved elements in the steel.
[viii] quasi-steady state has been established.

Problem 7 Solution

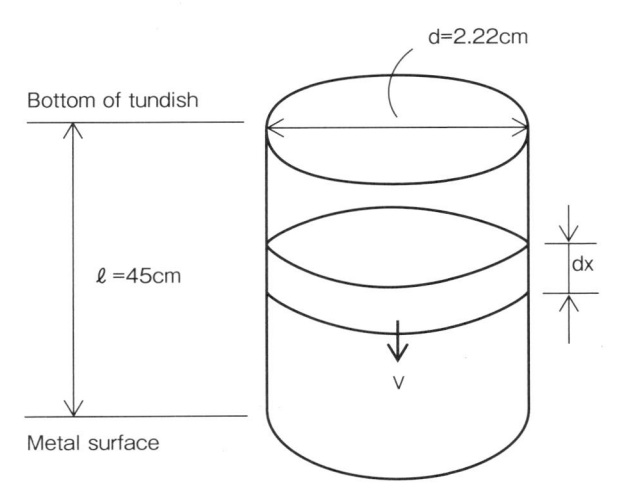

Fig. 6-7-2 Simplified casting stream from tundish to mold meniscus including controlled volume element thickness of dx.

図6-7-2に示す円柱状(直径d、長さℓ)の注入流れを想定する。その中間に厚みdxの要素体積を考える。

<Data>

$Q = 5.4$ kg/sec

$C_{0,0}^b = 4.5 \times 10^{-6}$ mole/cm^3

$\ell = 45$ cm

$\rho_{Fe} = 7.4$ g/cm^3

$D_O = 1.0$ cm^2/sec

$T_{air} = 770$℃

$P_{O_2} = 0.21$ atm

$R = 82.1$ cm^3·atm/mole·k

$K = 1.6$ mole/cm^3·atm

$d = 2.22$ cm

鋳造データは左図の通りである。

Q : 注入流量
$C_{0,0}^b$: 溶鋼中酸素の初期濃度(タンディッシュ出口における濃度)
ρ_{Fe} : 溶鋼の密度
D_O : 溶鋼中の酸素の拡散係数
T_{air} : 流れ周囲の温度
P_{O_2} : 大気の酸素分圧
R : 気体定数
K : 大気中酸素の溶鋼への溶融(吸収)反応の平衡定数

① mass balance (Molar balance)

Rate of oxygen input to the control volume is

$$v \cdot \left(\frac{d}{2}\right)^2 \cdot \pi \cdot C_o^b + \left(\frac{\dot{n}_o}{A}\right) \cdot d \cdot \pi \cdot dx \quad (6\text{-}7\text{-}2)$$

体積要素への酸素のインプット速度は(6-7-2)式、アウトプット速度は(6-7-3)式のようになる。

Rate of oxygen output from the control volume is

$$v \cdot \left(\frac{d}{2}\right)^2 \cdot \pi \cdot C_o^b + \frac{d}{dx}\left\{v \cdot \left(\frac{d}{2}\right)^2 \cdot \pi \cdot C_o^b\right\}dx \quad (6\text{-}7\text{-}3)$$

From eq. (6-7-2) and eq. (6-7-3)

(6-7-2)式と(6-7-3)式より、(6-7-5)式が得られる。

$$\frac{\dot{n}_o}{A} \cdot \pi d \cdot dx = \frac{d}{dx} C_o^b \cdot v \cdot \left(\frac{d}{2}\right)^2 \cdot \pi \cdot dx \quad (6\text{-}7\text{-}4)$$

$$\therefore \frac{\dot{n}_o}{A} = \frac{vd}{4}\frac{dC_o^b}{dx} \quad (6\text{-}7\text{-}5)$$

② molar flux

流れと気体間の界面における酸素分圧の分布を図6-7-3のように考える。なお、溶鋼中の酸素濃度は半径方向で均一とみなす。

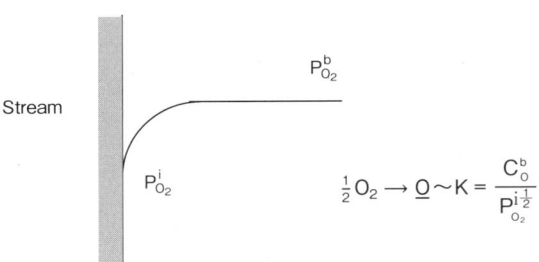

Stream $P_{O_2}^i$ $P_{O_2}^b$ $\quad \frac{1}{2}O_2 \rightarrow \underline{O} \sim K = \frac{C_o^b}{P_{O_2}^{i\,\frac{1}{2}}}$

Fig. 6-7-3 Distribution of oxygen near the stream
(Reaction at interface: Assuming the concentration of oxygenin in the stream is constant in radial direction)

③ stoickiometory

$$-\dot{n}_{O_2} = \frac{1}{2}\dot{n}_o \quad (6\text{-}7\text{-}6)$$

モル流束は(6-7-7)式のようになる。ここでk_oは界面気相側の酸素の物質移動係数である。

$$\frac{\dot{n}_o}{A} = 2k_o(C_{O_2}^b - C_{O_2}^i) = 2k_o\left(\frac{P_{O_2}^b}{RT} - \frac{P_{O_2}^i}{RT}\right) \quad (6\text{-}7\text{-}7)$$

C_{O_2} in bulk (P_{O_2} in bulk), C_{O_2} at interface (P_{O_2} at interface)

$$\therefore \frac{\dot{n}_o}{A} = 2k_o\left\{\frac{P_{O_2}^b}{RT} - \frac{(C_o^b)^2}{K^2 RT}\right\} \quad (6\text{-}7\text{-}8)$$

From eq. (6-7-5) and eq. (6-7-8)

(6-7-5)式と(6-7-8)式より(6-7-9)式が得られるが、これを(6-7-10)式のように変形し、xとC_o^bについて積分すると最終的に(6-7-14)式が得られる。

$$2k_o\left\{\frac{P_{O_2}^b}{RT} - \frac{(C_o^b)^2}{K^2 RT}\right\} = \frac{\pi d}{4}\frac{dC_o^b}{dx} \quad (6\text{-}7\text{-}9)$$

$$\therefore P_{O_2}^b - \frac{(C_o^b)^2}{K^2} = \frac{vdRT}{8k_o} \cdot \frac{dC_o^b}{dx} \quad (6\text{-}7\text{-}10)$$

$$\therefore \frac{8k_O}{vdRT}\int_0^\ell dx = \int_{C_{O,0}^b}^{C_{O,\ell}^b} \frac{1}{P_{O_2}^b - \frac{(C_O^b)^2}{K^2}} dC_O^b \quad (6\text{-}7\text{-}11)$$

$$\frac{8k_O}{vdRT}\int_0^\ell dx = \frac{1}{2(P_{O_2}^b)^{\frac{1}{2}}} \int_{C_{O,0}^b}^{C_{O,\ell}^b} \left\{ \frac{1}{(P_{O_2}^b)^{\frac{1}{2}} + \frac{C_O^b}{K}} + \frac{1}{(P_{O_2}^b)^{\frac{1}{2}} - \frac{C_O^b}{K}} \right\} dC_O^b$$

$$(6\text{-}7\text{-}12)$$

$$\frac{8k_O}{vdRT}\ell = \frac{K}{2(P_{O_2}^b)^{\frac{1}{2}}} \left[\ln\left\{ \frac{(P_{O_2}^b)^{\frac{1}{2}} + \frac{C_{O,\ell}^b}{K}}{(P_{O_2}^b)^{\frac{1}{2}} + \frac{C_{O,0}^b}{K}} \right\} - \ln\left\{ \frac{(P_{O_2}^b)^{\frac{1}{2}} - \frac{C_{O,\ell}^b}{K}}{(P_{O_2}^b)^{\frac{1}{2}} - \frac{C_{O,0}^b}{K}} \right\} \right]$$

$$(6\text{-}7\text{-}13)$$

$$\therefore \frac{16(P_{O_2}^b)^{\frac{1}{2}} \cdot k_O \cdot \ell}{vdRTK} = \ln\left\{ \frac{(P_{O_2}^b)^{\frac{1}{2}} + \frac{C_{O,l}^b}{K}}{(P_{O_2}^b)^{\frac{1}{2}} + \frac{C_{O,0}^b}{K}} \cdot \frac{(P_{O_2}^b)^{\frac{1}{2}} - \frac{C_{O,0}^b}{K}}{(P_{O_2}^b)^{\frac{1}{2}} - \frac{C_{O,l}^b}{K}} \right\} \quad (6\text{-}7\text{-}14)$$

where $k_O = 2\sqrt{\dfrac{D_O}{\pi t_e}} = 2\sqrt{\dfrac{D_O v}{\pi \ell}}$ \quad (6-7-15)

$$v = \frac{5.4 \times 10^3}{7.4 \times (\frac{2.22}{2})^2 \pi} = 188.62 \text{cm/sec}$$

$$k_O = 2\sqrt{\frac{1 \times 188.62}{\pi \times 45}} = 2.31 \text{cm/sec}$$

$$\ln\left\{ \frac{(0.21)^{\frac{1}{2}} + \frac{C_{O,l}^b}{1.6}}{(0.21)^{\frac{1}{2}} + \frac{4.5 \times 10^{-6}}{1.6}} \times \frac{(0.21)^{\frac{1}{2}} - \frac{4.5 \times 10^{-6}}{1.6}}{(0.21)^{\frac{1}{2}} - \frac{C_{O,l}^b}{1.6}} \right\} = $$

$$\frac{16 \times (0.21)^{\frac{1}{2}} \times 2.31 \times 45}{188.62 \times 2.22 \times 82.1 \times 1043 \times 1.6}$$

$$\therefore C_{O,\ell}^b = 9.38 \times 10^{-6}$$

Answer is,

$$4.50 \times 10^{-6} \frac{\text{mole}}{\text{cm}^3} \rightarrow 9.38 \times 10^{-6} \frac{\text{mole}}{\text{cm}^3}$$

(9.7ppm) \qquad (20.3ppm)

at tundish outlet \quad at the meniscus

溶鋼中の酸素濃度は9.7ppmから20.3ppmに増加すると予測される。

Problem 8
Manganese Loss During Steelmaking

F社は研究部門にメタルからスラグへのMnロスに関する調査を依頼した。そこで研究者らは実験装置を作成し1600℃において溶鋼からスラグへMnが移動する(損失する)実験を行った。

この移動はスラグ／メタル界面での(6-8-1)式で表される反応を伴っている。

この反応の平衡定数は(6-8-2)式で表される。

なお溶鋼中に、不活性ガスを吹き込むことによって強制対流を形成させた。

The Research Department of F company has been asked to investigate the rate of manganese loss from metal to slag during steelmaking processing. The research engineers have constructed a model, and have performed experiments in which manganese was transferred from liquid iron to a molten slag at 1600℃. This mass transfer was accompanied by a reaction at the interface

$$(FeO)_{slag} + [Mn]_{metal} = Fe + (MnO)_{slag} \quad (6\text{-}8\text{-}1)$$

for which the equilibrium constant is

$$K = \frac{(MnO)}{(FeO)[Mn]} = 2.41 \ (10^2) \quad (6\text{-}8\text{-}2)$$

where ( ) and [ ] are molar concentrations in the slag and metal phase respectively. Forced convection was provided by an ascending stream of inert gas bubbles.

この実験においてはFeOを14.4モル/ℓ含むスラグ、Mnを0.1モル/ℓ含む銑鉄が用いられた。

実験の進行に伴って鉄サンプルを採取しMn濃度の変化を調査した。表6-8-1にその結果を示す。

In one such experiment, the slag (35% CaO, 20% SiO_2, 35% FeO, 10% MgO) contained FeO at a concentration of 14.4

Table. 6-8-1 Change of Mn content in iron witgh time.

Time (min.)	[Mn] (mole ℓ$^{-1}$) × 10^3
2	87.1
4	72.4
6	60.2
8	52.5
10	44.6
12	36.3
14	33.1
16	27.6
18	24.0
20	19.5

mole ℓ^{-1}, while the concentration of Mn in the iron initially was 0.1 mole ℓ^{-1}. An analysis for manganese in samples of iron which were taken at various times gave the data listed in Table 6-8-1.

The volumes of the slag and metal phase were each 1.0 ℓ while the nominal area of the interface was 100 cm^2.

Calculate the overall mass transfer coefficient from these data, assuming that the transfer of FeO is not a rate determining step. Assess the relative importance of the Mn and MnO transfer coefficients to the overall mass transfer coefficient.

If the bubble frequency is 1 sec^{-1} and $D_{Mn} = 7(10^{-5})$ cm^2 sec^{-1}, $D_{MnO} = 10^{-6}$ cm^2 sec^{-1}, calculate the mass transfer coefficient from theory, and compare it with the value obtained experimentally.

実験におけるスラグ、メタルの体積はそれぞれ1ℓ、界面面積は100cm^{2}であった。

[1] これらのデータよりFeOの移動が律速段階でないと仮定して総括物質移動係数を求めよ。
[2] 総括物質移動係数に対するMnやMnOの移動係数が相対的に重要であることを確認せよ。
[3] 気泡の発生速度が1個/秒の場合、さらにD_{Mn}やD_{MnO}が左記のように与えられている場合、総括物質移動係数を求め実験データと比較せよ。

Problem 8 Solution

Concentration profile in the slag and metal system is shown in Fig. 6-8-1.

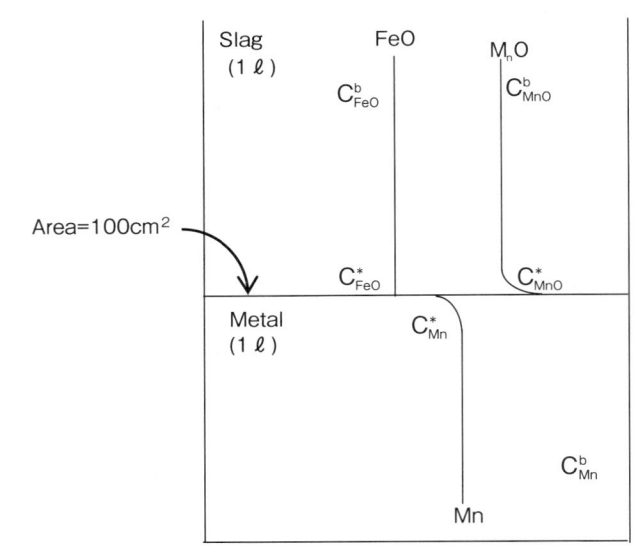

Fig. 6-8-1 Concentration profile in the slag and metal system.

slag/metal反応における濃度プロフィールを図6-8-1に示す。

a) 反応は瞬時に進行する
b) FeOの移動は反応律速ではない

という仮定の下、以下に数式で反応を記述する。

MnOはMnとFeOとの間に(6-8-3)式の平衡関係を有する。

FeOは律速過程でないため界面濃度とバルク濃度が等しいとみなし(6-8-4)式のようになる。

MnO、およびMnのモル流束の式は(6-8-5)、(6-8-6)式で表される。

これらのモル流束は等しいため(6-8-7)式が成立する。

<Assumptions>

a) The reactions take place instantaneously
b) Transport of FeO is not a rate controlling step.
  $(C_{FeO}^b = C_{FeO}^*)$

From the assumptions

$$C_{MnO}^* = K \cdot C_{FeO}^* \cdot C_{Mn}^* \qquad (6\text{-}8\text{-}3)$$

$$= K \cdot C_{FeO}^b \cdot C_{Mn}^* \qquad (6\text{-}8\text{-}4)$$

$$-\frac{\dot{n}_{MnO}}{A} = k_{MnO}(C_{MnO}^b - C_{MnO}^*) \qquad (6\text{-}8\text{-}5)$$

$$\frac{\dot{n}_{Mn}}{A} = k_{Mn}(C_{Mn}^b - C_{Mn}^*) \qquad (6\text{-}8\text{-}6)$$

$$\dot{n}_{MnO} = \dot{n}_{Mn} \qquad (6\text{-}8\text{-}7)$$

From eqs. (6-8-4)～(6-8-7)

$$C_{Mn}^* = \frac{k_{MnO}C_{MnO}^b + k_{Mn}C_{Mn}^b}{k_{Mn} + KC_{FeO}^b k_{MnO}} \quad (6\text{-}8\text{-}8)$$

(6-8-4)～(6-8-7)式より(6-8-8)式を得る。

From eqs. (6-8-6) and (6-8-8)

$$\frac{\dot{n}_{Mn}}{A} = k_{Mn}\left(C_{Mn}^b - \frac{k_{MnO}C_{MnO}^b + k_{Mn}C_{Mn}^b}{k_{Mn} + KC_{FeO}^b k_{MnO}}\right) \quad (6\text{-}8\text{-}9)$$

$$= \frac{1}{\frac{1}{k_{Mn}} + \frac{1}{k_{MnO}KC_{FeO}^b}} \left(C_{Mn}^b - \frac{C_{MnO}^b}{KC_{FeO}^b}\right) \quad (6\text{-}8\text{-}10)$$

$$= k_{ov}\left(C_{Mn}^b - \frac{C_{MnO}^b}{KC_{FeO}^b}\right) \quad (6\text{-}8\text{-}11)$$

(6-8-6)式に(6-8-8)式を代入すると(6-8-11)式が得られる。
ここでk_{ov}は総括物質移動係数である。

Taking a molar balance,

$$V_s(C_{MnO}^b - C_{MnO,0}^b) = V_M(C_{Mn,0}^b - C_{Mn}^b) \quad (6\text{-}8\text{-}12)$$

$$\therefore C_{MnO}^b = \frac{V_M}{V_S}(C_{Mn,0}^b - C_{Mn}^b) + C_{MnO,0}^b \quad (6\text{-}8\text{-}13)$$

スラグ／メタル間のモルバランスの式は(6-8-12)式となる。
(6-8-11)式と(6-8-13)式から、(6-8-14)、(6-8-15)式を得る。

$$\therefore \frac{\dot{n}_{Mn}}{A} = k_{ov}\left\{C_{Mn}^b - \frac{1}{KC_{FeO}^b}\frac{V_M}{V_S}(C_{Mn,0}^b - C_{Mn}^b) - \frac{C_{MnO,0}^b}{KC_{FeO}^b}\right\}$$

$$= k_{ov}\left\{C_{Mn}^b\left(1 + \frac{V_M}{V_S}\frac{1}{KC_{FeO}^b}\right) - \frac{V_M C_{Mn,0}^b}{V_S KC_{FeO}^b} - \frac{C_{MnO,0}^b}{KC_{FeO}^b}\right\}$$

$$= k_{ov}(\alpha C_{Mn}^b + \beta) \quad (6\text{-}8\text{-}14)$$

where

$$\alpha = 1 + \frac{V_M}{V_S}\cdot\frac{1}{KC_{FeO}^b}, \quad \beta = -\frac{V_M C_{Mn,0}^b}{V_S KC_{FeO}^b} - \frac{C_{MnO,0}^b}{KC_{FeO}^b} \quad (6\text{-}8\text{-}15)$$

$$-\frac{\dot{n}_{Mn}}{A}A = V_M \frac{dC_{Mn}^b}{dt} \quad (6\text{-}8\text{-}16)$$

$$\therefore \frac{dC_{Mn}^b}{dt} = -\frac{A}{V_M}k_{ov}(\alpha C_{Mn}^b + \beta) \quad (6\text{-}8\text{-}17)$$

$$\int_{C_{Mn,0}^b}^{C_{Mn}^b}\frac{dC_{Mn}^b}{\alpha C_{Mn}^b + \beta} = -\frac{A}{V_M}k_{ov}\int_0^t dt \quad (6\text{-}8\text{-}18)$$

$$\therefore \frac{1}{\alpha}\ln\left(\frac{\alpha C_{Mn}^b + \beta}{\alpha C_{Mn,0}^b + \beta}\right) = -\frac{A}{V_M}k_{ov}t \quad (6\text{-}8\text{-}19)$$

(6-8-14)、(6-8-16)式より(6-8-17)式が得られるが、この式をC_{Mn}^bとtについて積分して(6-8-19)式を得る。
これより$\left\{-\frac{V_M}{A}\frac{1}{\alpha}\ln\left(\frac{\alpha C_{Mn}^b + \beta}{\alpha C_{Mn,0}^b + \beta}\right)\right\}$ をtに対してプロットすればその傾きが総括物質移動係数k_{ov}となる。

So if we plot $\left\{-\frac{V_M}{A}\frac{1}{\alpha}\ln\left(\frac{\alpha C_{Mn}^b + \beta}{\alpha C_{Mn,0}^b + \beta}\right)\right\}$ to t, the gradient of that line gives a value of k_{ov}.

Table. 6-8-2 Relation of t to $\left\{-\dfrac{V_M}{A}\dfrac{1}{\alpha}\ln\left(\dfrac{\alpha C_{Mn}^b + \beta}{\alpha C_{Mn,0}^b + \beta}\right)\right\}$

t	C_{Mn}^b	$\left\{-\dfrac{V_M}{A}\dfrac{1}{\alpha}\ln\left(\dfrac{\alpha C_{Mn}^b + \beta}{\alpha C_{Mn,0}^b + \beta}\right)\right\}$
2	87.1×10^{-3}	1.381
4	72.4×10^{-3}	3.230
6	60.2×10^{-3}	5.075
8	52.5×10^{-3}	6.444
10	44.6×10^{-3}	8.076
12	36.3×10^{-3}	10.136
14	33.1×10^{-3}	11.059
16	27.6×10^{-3}	12.877
18	24.0×10^{-3}	14.276
20	19.5×10^{-3}	16.354

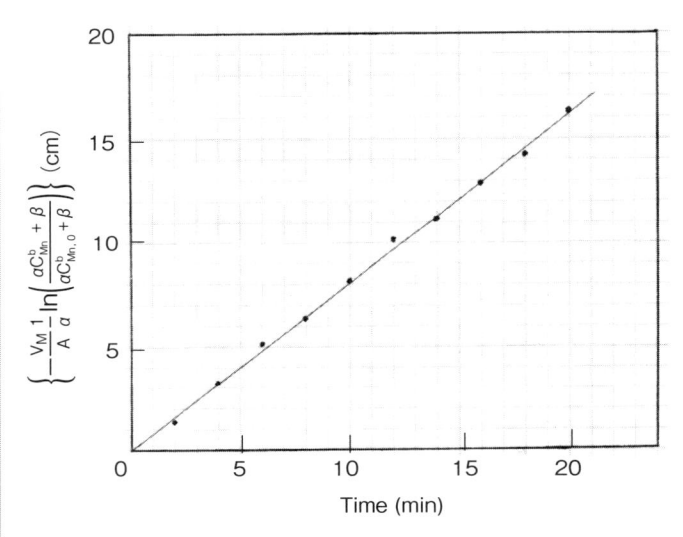

Fig. 6-8-2 Relation between t and $\left\{-\dfrac{V_M}{A}\dfrac{1}{\alpha}\ln\left(\dfrac{\alpha C_{Mn}^b + \beta}{\alpha C_{Mn,0}^b + \beta}\right)\right\}$.

図6-8-2よりk_{OV} は 0.013(cm/sec) と求まる．

一方、浸透説よりk_{Mn}、k_{MnO}は (6-8-20)、(6-8-21)式から求まり、(6-8-22)式で定義される総括物質移動係数は9.42×10^{-3}(cm/sec)となる。

From this figure, $k_{OV} = 0.013$ (cm/sec).

On the other hand, from the penetration theory,

$$k_{Mn} = 2\sqrt{\dfrac{D_{Mn}\cdot f}{\pi}} = 9.44 \times 10^{-3} \text{ (cm/sec)} \qquad (6\text{-}8\text{-}20)$$

$$k_{MnO} = 2\sqrt{\dfrac{D_{MnO}\cdot f}{\pi}} = 1.13 \times 10^{-3} \text{ (cm/sec)} \qquad (6\text{-}8\text{-}21)$$

$$\therefore k_{OV} = \cfrac{1}{\cfrac{1}{k_{Mn}} + \cfrac{1}{k_{MnO} \cdot K \cdot C_{FeO}^{b}}} \qquad (6\text{-}8\text{-}22)$$

$$= 0.0094 \quad (\text{cm/sec})$$

There is certain discrepancy between measured k_{OV} and calculated (theoretical) k_{OV}.

したがって測定されたk_{OV}と浸透説で計算されるk_{OV}は必ずしも一致しない事になる。

Additional Short Discussions in Metallurgical Mass Transfer

A-1. Decarburization in Q-BOP

In the Q-BOP steelmaking process, oxygen enriched gas is introduced via a tuyere at the bottom of the vessel as shown in Fig. A-1 and decarburization takes place at gas bubble/liquid metal interfaces by the following reaction.

$$[C] + 1/2\, O_2\,(g) \rightarrow CO\,(g) \qquad (A\text{-}1)$$

At the steelmaking temperatures of 1600℃ this reaction is virtually instantaneous, has a large equilibrium constant and is mass transfer controlled. A study of the kinetics of the process in a 30t pilot plant vessel gave the results shown in Fig. A-2, for the conditions given below.

Q-BOPとはQ (Quality etc.) -Basic Oxygen Processの略で炉底部に設けた羽口から酸素やフラックスを吹き込んで精錬するプロセスで、強力な攪拌エネルギーを特徴としている（図A-1参照）。

1600℃以上の精錬温度では、(A-1)式の脱炭反応は非常に早く進行し、物質移動律速であると考えられる。30トン規模のパイロットプラント実験の結果を図A-2に示す。

右記の条件をもとに、(1)図A-2の領域Ⅰ、Ⅱのそれぞれの律速過程を示すと共にⅠからⅡに遷移する際のC濃度を求めよ。また、(2)C濃度が1%の鋼を処理する際の初期脱炭速度を求めよ。

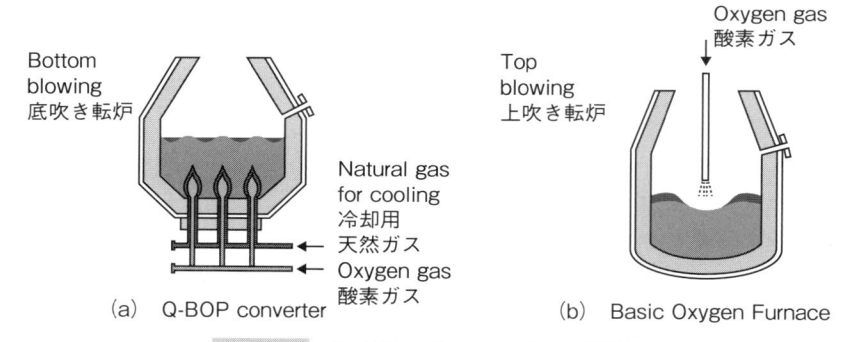

Fig. A-1 Q-BOP and conventional BOF[5]

(a) Q-BOP converter
(b) Basic Oxygen Furnace

Liquid phase mass transfer coefficient	= 0.03 cm/s
Gas phase mass transfer coefficient	= 4.50 cm/s
Pressure of oxygen in gas bubbles	= 1 atm
Bubble/metal interfacial area	= 2.04(10^6) cm^2
Density of steel	= 7 g/cm^3
Molecular weight of carbon	= 12 g/mole
Gas constant, R	= 82.1 cm^{3}atm/mole K

(1) Determine the rate controlling mechanisms in each of the regions shown in Fig. A-2 and calculate the carbon content at which the transition occurs, and compare with the observed result.

(2) Calculate from mass transfer theory the initial decarburization rate for a batch carbon content of 1 wt %.

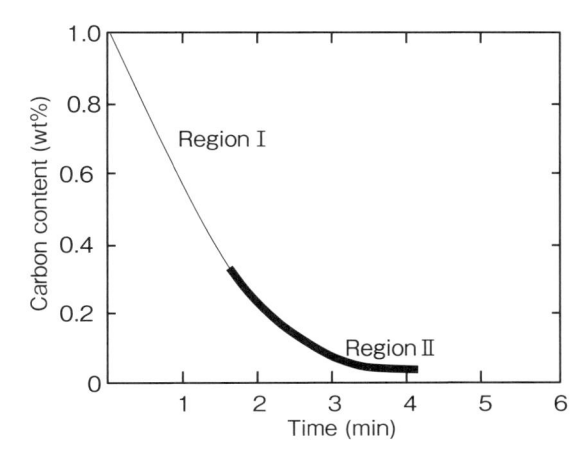

Fig. A-2 Decarburization of molten steel in Q-BOP process.

A-2. Evolution of CO in the refining of low carbon ferro-chrome alloy

低炭フェロクロム合金や高クロム鋼は、一般的には電気炉を使い、クロム鉱石に還元剤としてコークス、フラックスとして硅石、石灰石等を加え溶融・還元する事によって製造される。

ここで重要なポイントはクロムの酸化を抑えつつカーボンの酸化を最小限にすることである。

Fe-Cr-C合金の脱炭機構を調査するために溶融Fe-Cr-C合金のプール中に柱状Cr_2O_3を浸漬・回転させ(A-2)式で表されるCOガスの発生量を測定した。ここで(A-2)式の反応は図A-3に示すようなステップで進み、(A)、(D)の反応は非常に速く、(B)、(C)のCやCrの移動は律速段階とはならないと仮定し、(A-3)式の形でCOガスの発生速度を予測する式を導きなさい。

In the production of low carbon ferro-chrome alloys and high chromium steels, an important goal is the oxidation of carbon to very low values without excessive simultaneous oxidation of chromium. A study was conducted on the mechanism of decarburization of Fe-Cr-C alloys by rotating a partially submerged cylinder of Cr_2O_3 in a pool of Fe-Cr-C alloy and measuring the rate of CO evolution, according to the following reaction.

$$Cr_2O_3 + 3C \rightarrow 2Cr + 3CO \text{ (g)} \qquad (A-2)$$

It has been postulated that reaction (A-2) takes place via a series of steps schematically shown in Fig. A-3. It is proposed that reactions (A) and (D) in Fig. A-3 are virtually instantaneous, and the transport of C and Cr in steps (B) and (C) are not rate controlling. To test this postulate derive a differential equation for predicting the volume of CO(g) evolved with time of the following form,

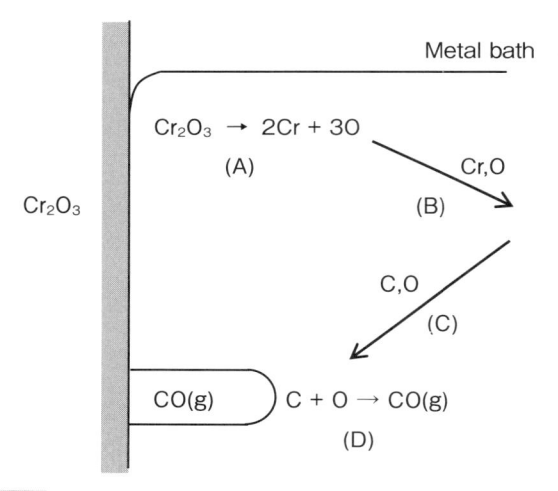

$$\frac{dV_{CO}}{dt} = f(V_{CO}, K_{(A)}, K_{(D)}, k_{O(B)}, k_{O(C)}, C_{Cr}, C_C, A_s, A_g, P_{CO}) \qquad (A-3)$$

Fig. A-3 Chemical reactions and mass transfer in the metal bath of low-carbon ferro-chrome alloy.

which could be utilized to calculate V_{CO} vs. t and compared with measurements. Do not attempt to solve the equation.

V_{CO} = Volume of CO evolved at a given time at STP
$K_{(A)}$ = Equilibrium constant for reaction (A)
$K_{(D)}$ = Equilibrium constant for reaction (D)
$k_{O(B)}$ = Mass transfer coefficient for oxygen in step (B)
$k_{O(C)}$ = Mass transfer coefficient for oxygen in step (C)
C_{Cr_i}, C_{C_i} = Initial concentration of Cr and C in the bath
A_s = Cr_2O_3/metal interfacial area
A_g = Metal/CO bubble interfacial area
P_{CO} = Pressure in CO bubble

あとがき

　反応プロセス工学という講義を担当することになって、適切な教材や素材を探そうと資料を整理している際、若いころUBCで学んだ物質移動に関する英語の講義ノートを偶然見つけた。ページをめくっていくにつれBrimacombe 教授やSamarasekera教授らの講義の様子が懐かしく思い出されると共に、ふと教える立場に戻った時、講義内容が基礎から応用まで上手く構成されていることに改めて気づかされた。それから数年間、このノートをベースに講義を行ってきたが、その都度気付いた修正や追補もほぼ整い、テキストにまとめることができないかと考えるようになった。Samarasekera先生に相談したところ、講義資料については、思うように活用するようにと、ご快諾頂いた。英文は当時の授業や宿題の雰囲気を伝えるべく、板書の写しや聞き取った表現に近い形とし、説明が不足していると思われる個所は和文で補うようにした。そのため、簡略化された部分や旧単位系などわかりにくい個所も多々あると思われるがご容赦いただきたい。

　最後に、J. Keith Brimacombe教授と、Indira V. Samarasekera教授に改めて感謝いたします。
　また、非常に読み難い手書きのノートや図表から電子ファイルを作成いただきました事務補佐員の藤原照美氏、度重なる構成の変更にも忍耐強く対応いただきました大阪大学出版会の栗原佐智子氏に感謝いたします。

2015年7月30日

著者

References

1) R. B. Bird, W. E. Stewart, E. N. Lightfoot, "Transport Phenomena / Revised 2nd Edition", John Wiley & Sons, Inc., (2007).
2) 水科篤郎, 荻野文丸, 『輸送現象』, 産業図書, (1981).
3) J. Szekely, N. J. Themelis, "Rate Phenomena in Process Metallurgy", John Willey & Sons, Inc., (1971).
4) 谷口尚司, 八木順一郎, 『材料工学のための移動現象論』, 東北大学出版会, (2011).
5) 「鉄ができるまで / Making of Iron & Steel」, (一社)日本鉄鋼連盟, (2015).
6) 講座・現代の金属学 製錬編編集委員会, 『講座・現代の金属学 製錬編第2巻 非鉄金属精錬』, 社団法人日本金属学会, (1980).
7) P. Atkins, J. Paula, "Physical Chemistry 8th Edition", Oxford University Press, (2006).
8) J. K. Brimacombe, E. S. Stratigakos, P. Tarassoff; *Metall. Trans.*, (1974), 763-71.
9) N. A. Warner, Kinetics of continuous vacuum dezincing of lead, in *Advances in Extractive Metallurgy: Proceedings of a symposium organized by the Institution of Mining and Metallurgy, held in London from 17 to 20 April, 1967*, Institution of Mining and Metallurgy: London, (1968), 317-32.

Index

A–Z

accommodation coefficient(s)　71
activation energy　8, 9
anode furnace　2, 57
anode(s)　91
barometric leg　65
basic oxygen furnace (BOF)　2, 83
batch process　41, 55
blast furnace (BF)　1
blister copper　57
bottom browing　124
cathode(s)　91, 92, 94, 95, 96
chemical reaction　7, 25
CO bubble　48
condensation　67, 70
conductance　18, 33
consecutive control　63
continuous casting　98, 112
continuous process　55
convection　11
copper converter/(anode furnace)　2
decarburization (rates)　105, 111, 124
diffusion coefficient　11, 33, 74
diffusivity　6
diffusion boundary layer　19
distillation　69
driving force　18, 33
electrolyte　91
electrowinning　91
equilibrium　9
equimolar counter diffusion　19
evaporation　16, 66, 67, 89
ferro-chrome alloy　126
Fick's 1st law　6, 11, 12
Fick's 2nd law　29
film theory　34, 95
flux　6
—— equation　17, 20, 22, 25
Fourier's law　7
fuel vapor　16
fume(s)　83, 84, 85

heat flux　7
heat transfer　33
heterogeneous reaction　19
homogeneous chemical reaction　24, 86
Imperial Smelting furnace (process)　3, 65, 74
interdiffusion coefficient　12, 15
iron　1, 105, 106, 108
iron vapor　83, 85
irreversibility　86
jet O_2　2
kinetic viscosity　7
Kirkendall effect　12
L'Hopital's rule　19
limiting current density　93
linear thermodynamics　44
local equilibrium　41, 45, 49
log mean mole fraction　19
manganese loss　118
mass balance　6, 16, 20, 22, 25
mass flux　6
mass transfer　33
mass transfer coefficient　33
mass transfer number　58, 61
maximum current density　92
molar balance　41, 44, 49, 52
molar concentration　11
molar density　12
mole(molar) flux　6, 11, 41, 42, 45, 49, 52, 53
molar fraction　12
mole center velocity　13
molecular diffusion　11
momentum flux　7
Newton's law　7
nonlinear thermodynamics　48
non-equimolar counter diffusion　22
Ohm's law　33
open pour　112, 113
overall mass transfer coefficient　46, 119
overall rate of decarburization

106
Pb　3
penetration theory　35, 37, 99, 113, 122
plug flow　66, 114
poling　57, 64
Q-BOP　124
rate of chemical reaction　8
rate controlling　47, 51, 52, 53, 57
Reynolds number　33
roasting　79
role similarity　33
Schmidt number　33
shear stress　7
Sherwood number　33
slag　1, 2, 3, 105, 106, 108
splash condenser　65
stagnant medium　16
steady state diffusion (Fick's 1st law)　11
steel　2, 4, 98, 99, 102, 112, 113, 124, 125
steelmaking　118, 124
stoichiometry　10, 20, 22, 25, 41, 43, 45, 49, 52, 53
thermal diffusivity　7
top blowing　124
topochemical (reaction)　79, 81
tundish　112
tungsten pellets　98
unsteady state diffusion　29
vacuum dezincing　65
volume velocity　13
zinc　65, 66, 74, 91, 94
Zn　3, 65, 66, 74, 94

あ 行

亜鉛　3, 67, 74, 91, 94
陰極　91, 94
Imperial Smelting 炉（亜鉛）　3, 65
Imperial Smelting 炉設備　5
運動量流束　7
液体燃料　16

オープン注入　112, 113

か 行

化学反応速度　8
化学量論式　45
化学量論的関係(化学量論性)　10, 20, 22, 49, 52, 53
活性化エネルギー　9
拡散係数　12
凝集　67, 70
境膜理論　34, 95
Kirkendall効果　12
均一(化学)反応　24, 86
駆動力　33
限界電流密度　92
高炉　1
高炉設備　4
コンダクタンス　33

さ 行

最大電流密度　92
蒸発　16, 66, 67, 89
蒸留　69
酸化鉄　1
CO気泡　48
浸透理論　35, 37
　　──説　99, 113, 122
スプラッシュコンデンサー　65
スラグ　1
銑鉄　1, 2
粗銅　57
遷移律速　63
全モル濃度　12
総括脱炭速度　106
総括物質移動係数　45, 46, 119

相互拡散係数　12

た 行

大気脚　65
対数平均　18
体積流れ　13
対流　11
脱炭(速度)　105, 111, 124
タンディッシュ　112
タングステンペレット　98
適応係数　71
鉄蒸気　83, 85
電解質溶液　91
電解析出　91
転炉　83
　　──(鋼)　2
　　──(銅)　2
　　──設備　4
銅　2
トポケミカル(反応)　79, 81
等モル相互拡散　19

な 行

鉛　3
熱流束　7
濃度遷移境界層　19
Newtonの法則　7

は 行

焙焼　79
バッチプロセス　41
反応平衡　41
反応速度定数　8
Peirce-Smith 転炉設備　5
非等モル相互拡散　22, 81

Fickの第一法則　6, 11
Fickの第二法則　32
Fourierの法則　7, 11
不可逆反応　86
不均一反応　19
物質移動係数　33
物質移動数　58, 61
物質収支の式　20, 59
物質流束　6
フェロクロム合金　126
フューム　83, 84, 85
プラグフロー　55, 66, 114
分子拡散　11
平衡状態　9
ポーリング　57

ま 行

マスバランス　6, 22, 49
Mnロス　118
モルバランス　41
モル分率　12
モル流束　12, 41, 42, 45, 49, 52, 53

や 行

陽極　91
溶鋼　2, 98, 99, 113, 114, 115, 118

ら 行

ランス　2
律速(段階)　47, 51, 52, 53, 57
流束　6
流束の式　20, 22, 25
連続真空亜鉛分離プロセス　65
連続鋳造　98, 112
連続プロセス　55

著者紹介

竹内 栄一（たけうち えいいち）
1977年　九州大学大学院工学研究科修了、新日本製鐵（株）
1982年　ブリティッシュコロンビア大学留学
1984年　Ph.D.取得後、新日本製鐵（株）
2011年　大阪大学大学院工学研究科教授（現職）

企業においては連続鋳造プロセスへの電磁力適用の研究開発、大学においてはプロセス反応工学分野の教育・基礎研究を中心に活動。
主な受賞歴としてHenry Marion Howe Medal/American Society for Metals（1986年）、John Chipman Award/Iron and Steel Society（1987年）、Best Paper Award/Canadian Institute of Mining, Metallurgy and Petroleum（2001年）、功績賞/日本金属学会（1996年）、西山記念賞／日本鋼鉄協会（2003年）など。

田中 敏宏（たなか としひろ）
1985年　大阪大学大学院工学研究科修了（工学博士）、大阪大学助手
1989年　ドイツ・アーヘン工科大学・理論冶金研究所・フンボルト財団研究員として1年間滞在
1995年　大阪大学工学部助教授
2002年　大阪大学大学院工学研究科教授（現職）

高温材料プロセスにかかわる界面現象・材料物理化学の分野を中心に活動。
主な受賞歴として、俵論文賞/日本鉄鋼協会（1984年）、本多記念研究奨励賞/本多記念会（1986年）、西山記念賞/日本鉄鋼協会（1996年）、功績賞/日本金属学会（2001年）、論文賞/日本金属学会（2004年）、学術功績賞/日本鉄鋼協会（2007年）など。

高温材料プロセスにおける
物質移動の基礎とケーススタディー
Metallurgical Mass Transfer — *from a lecture note of Professors Brimacombe and Samarasekera*

2015年10月1日　　初版第1刷発行　　　　［検印廃止］

著　者　竹内栄一，田中敏宏

発行所　大阪大学出版会
　　　　代表者　三成賢次
　　　　〒565-0871　大阪府吹田市山田丘2-7
　　　　　　　　　　大阪大学ウエストフロント
　　　　電話（代表）06-6877-1614
　　　　FAX　　　　06-6877-1617
　　　　URL　　　　http://www.osaka-up.or.jp
印刷・製本　亜細亜印刷株式会社

ⒸEiich Takeuchi & Toshihiro Tanaka 2015　　　Printed in Japan
ISBN 978-4-87259-503-1 C3057

Ⓡ〈日本複写権センター委託出版物〉
本書を無断で複写複製（コピー）することは、著作憲法上の例外を除き、禁じられています。本書をコピーされる場合は、事前に日本複写権センター（JRRC）の許諾を受けてください。